Edward G. Begle
1914-1978

CRITICAL VARIABLES IN MATHEMATICS EDUCATION:
Findings from a Survey of the Empirical Literature

E. G. Begle

Stanford University

Published by the Mathematical Association of America
and the National Council of Teachers of Mathematics

Washington, D.C.

This survey was conducted without financial assistance
from any agency of the U.S. Government.

CRITICAL VARIABLES
IN MATHEMATICS EDUCATION:

CONTENTS

Editors' Preface

On 2 March 1978, after battling emphysema and its complications
for over two years, Edward Griffith Begle died in Palo Alto, California,
at the age of 63, and an era in mathematics education came to a close.
At a memorial service two days later in the Stanford Memorial Church,
four of his colleagues and friends eulogized him as a human being and
as a scholar. Perhaps the most touching eulogy, however, came at the
end of the service and was unspoken. A bagpiper played several hymns
at the rear of the church and then, still playing, turned and walked
slowly out the door, the music fading gradually away. In the silence
that followed, one could feel how aptly the bagpipe music had expressed
the dignity, the integrity, and the warmth of Ed Begle.

If any of Ed's colleagues assumed that his death meant the end of
his innumerable contributions to the field of mathematics education,
they underestimated him. The manuscript of this book had been completed only weeks before and was still being given final touches by
its author, who would not let the tightening grip of his illness stop
him from completing a job he had set out to do.

Historians of education will probably consider Ed Begle a curriculum developer, citing his directorship of the School Mathematics Study
Group as his main accomplishment. But they will thereby miss much of his
work in mathematics education. As a mathematics educator, Ed Begle was a
virtuoso. He began his career as a mathematician, turned to curriculum

development with the advent of SMSG, and gradually broadened his
concern to include all realms of research in mathematics education.
This book is the capstone of his career as mathematician and educator.
Its publication jointly by the National Council of Teachers of Mathematics and the Mathematical Association of America is eminently fitting
since Ed made many of his contributions to mathematics education
through these two organizations.

From the earliest days of SMSG, Ed Begle had been concerned that
curriculum development in mathematics education be provided with a
foundation of empirical research. He initiated the National Longitudinal Study of Mathematical Abilities to investigate various factors
that might affect the learning of mathematics. When Ed moved to Stanford
in the autumn of 1961, he began a series of seminars on research for
his students and colleagues that continued up to his death. At first
these seminars explored questions raised by NLSMA, but later they
ranged widely as Ed's interest and energy yielded a variety of topics
for exploration.

The activities of NLSMA and the seminars, as well as other activities of SMSG, soon led to the establishment of a library of books,
microfilms, and articles on mathematics education that eventually became one of the best and most extensive of its kind. The bulk of the
collection consisted of reports or reviews of empirical research and
formed the basic set of material used in writing this book.

Ed's experience in NLSMA convinced him that mathematics education
should develop a stronger empirical base for its activities. He tried
to persuade the SMSG Advisory Board to sponsor research as well as
curriculum development, but he was not successful. When the "second
round" of SMSG curriculum development began in the late 1960's, it
was more systematic and deliberate than the first round had been, but
it was not research-based.

Ed, however, was undaunted. At the First International Congress
on Mathematical Education in Lyon in 1969, in an address entitled
"The Role of Research in the Improvement of Mathematics Education,"
he stated his position:

    I see little hope for any further substantial
improvements in mathematics education until we turn
mathematics education into an experimental science,

until we abandon our reliance on philosophical discussion based on dubious assumptions and instead
follow a carefully constructed pattern of observation
and speculation, the pattern so successfully employed
by the physical and natural scientists.

We need to follow the procedures used by our
colleagues in physics, chemistry, biology, etc. in
order to build up a theory of mathematics education
. . . . We need to start with extensive, careful,
empirical observations of mathematics learning. Any
regularities noted in these observations will lead to
the formulation of hypotheses. These hypotheses can
then be checked against further observations, and
refined and sharpened, and so on. To slight either
the empirical observations or the theory building
would be folly. They must be intertwined at all times.

As Ed put it, the curriculum reform activities of SMSG had helped
bring under control the problem of teaching better mathematics; not
under control, and open to research, was the problem of teaching mathematics better. He, together with his students and colleagues, began
a series of reviews of the empirical literature on such topics as
mathematics laboratories and individually prescribed instruction. He
turned to the NLSMA data pool for information on teaching effectiveness
and on the prediction of student achievement. He and his students began
studies on the effects of manipulating instructional variables such as
the extent of practice and the use of review. A serious weakness of
much research on teaching methods has been associated with the mani-
fold dimensions along which any two methods are likely to vary: If
the two methods differ in effects, the cause cannot ordinarily be
determined. Ed and his students attempted to deal with this problem
by divising some clearly structured teaching units that would permit
systematic variation along one dimension at a time. They then used
these units in various research studies.

In 1972, when SMSG had concluded its work, Ed formed the Stanford
Mathematics Education Study Group (SMESG) to continue the literature
reviews and empirical studies that he saw as necessary to the establishment of mathematics education as an experimental science. He also
continued his seminars for graduate students in mathematics education
at Stanford, and it was probably in these seminars--in the process of

acquainting students with the work that had been done in their field--
that he first saw the potential value of compiling reviews of the
literature on all facets of mathematics education. This book will be
valuable to many people for the information it contains, but no one
will profit from it more than beginning graduate students in mathematics education.

By the fall of 1976, Ed had prepared dozens of short bibliographies
for use in his seminars. Each bibliography contained references on a
given topic in mathematics education; the references were selected to
illustrate the nature and variety of research on the topic and served
as stimuli for discussion in the seminar. He had also assembled and
classified almost seven thousand index cards containing abstracts and
bibliographic information on the studies he had located, most of them
obtained from the search he describes in the Preface.

During the months of recuperation from his hospitalization in
December 1976, Ed began working on the manuscript for this book.
First drafts of the first six chapters were sent to some of his colleagues throughout the country for their comments and criticisms during
the summer of 1977. By December 1977, the remaining chapters had been
written, the earlier ones revised, and a more polished draft of the
entire manuscript prepared for circulation to his colleagues. A few
days later he entered the hospital for the last time.

In the weeks after Ed's death, several of his colleagues conferred
as to how to arrange for publication of the manuscript. The manuscript
was, in some respects, unfinished, for time had not permitted Ed to respond to all the comments and suggestions of his colleagues. Furthermore,
it was timely in the sense that the literature under review did not go
beyond 1976.

Several alternatives were considered, but joint MAA-NCTM publication and distribution were deemed best. Comments and suggestions were
solicited from the 13 colleagues who had been asked to review the final
draft. These comments and suggestions were sent to the University of
Georgia, where we attempted to incorporate as many as possible of them
into the manuscript without destroying its fabric. Everyone who has
discussed the publication of the manuscript with us has agreed that

it should be published essentially as Ed left it, and our changes have been limited to correcting errors and attempting to reduce ambiguity and inconsistency. With the cooperation of the officers and the publication committees of the MAA and the NCTM, the manuscript was reviewed and approved in record time--a tribute to the profession's esteem and affection for the author. Special thanks for their assistance are due Glenadine Gibb, past-president of the NCTM, Leonard Gillman, treasurer of the MAA, and Henry Pollak, past-president of the MAA.

This book is like one of those photographs taken from a tower or mountaintop showing a 360° view of the terrain. It gives a panorama, an overview of the landscape of research in mathematics education. It does not contain fine detail. The reader who goes looking for information on a specific topic may be disappointed. "What good does it do me," the reader may grumble, "to know that Begle found 17 references on class climate? I want to know what they were and what they said." For such information, the reader will have to look elsewhere. (The Preface suggests some sources, and a document containing the bibliographic information from the index cards used in preparing the manuscript is being entered into the ERIC system.) The purpose of this book is to give the reader an idea of the attention researchers have paid to a given topic, and wherever possible it contains Ed Begle's assessment of what has been done and the prospects for further work. It yields what its title suggests: a survey.

A photograph depends upon the lens used in taking it. This book was written by one of the most eminent mathematics educators of our time. It is not simply a tally of references assembled by a bibliographer. It bears Ed Begle's stamp. Almost every page is illuminated by his observations and opinions. One can argue--and several reviewers have--that the first chapter's discussion of mathematics and mathematical objects is discordant with the rest of the book, but that is how he left it. One can complain that in reviewing the literature on, say, ability grouping, he has overlooked contrary arguments and evidence. One can puzzle over his reasons for concluding that on one topic further work is needed, whereas on another topic further work would be fruitless, when both topics appear equally unexplored. The careful reader will finish the book with a long list of questions that might have been asked of the author. No one would have been happier to discuss those questions than Ed Begle.

While preparing the manuscript of this book for publication over the past few months, we have used a copy of Ed's index card file in checking various references. Although both of us were already reasonably familiar with the work of Ed and his students, we were impressed by the great range of this work. The set of index cards for almost every major topic in the book contains references by Ed or by his students. We know of no other mathematics educator with such catholic interests and accomplishments in research. We consider this book a unique contribution to our field.

The book ends on a gloomy note. Ed was depressed that little new knowledge about mathematics education emerged in the last decade, and that we still lack theoretical structures to support our research. Depressed, he says, but not despondent since the empirical base for further progress is there in the literature, waiting to be uncovered. No one who has looked closely at the research literature in mathematics education can be sanguine about its present state. A mild depression may be the only appropriate response. Anyone who knew Ed Begle, however, knows that he would not have let such a reaction stop him from going forward, and we who are in his debt should not let it stop us.

James W. Wilson
Jeremy Kilpatrick

Athens, Georgia
October 1978

Preface

The original reason for undertaking this survey was to provide
some guidance to those interested in conducting comprehensive reviews
of the factual information which exists about the effects of various
variables on student learning of mathematics. A few such reviews have
been carried out and their findings have turned out to be remarkably
useful. But there are very many potentially important variables which
might affect mathematics learning and it was not clear which ones
should be investigated next. I felt that a survey of all of them
might suggest some reasonable priorities and thus indicate which of
these variables are the most critical.

However, as the survey progressed, it became clear that it could
be useful to the much wider audience of all those who are directly
concerned with mathematics education and, in particular, its improvement: mathematics educators, classroom teachers, mathematicians,
school administrative personnel, teachers of teachers, textbook authors,
etc. Indeed, some parts of the survey may be of interest to the lay
public and especially to parents.

It is not generally realized by most of these individuals that a
very large number of experiments and other empirical studies on mathematics education have been carried out. But further improvements in
mathematics education require that knowledge about the findings of these
experiments and studies become widespread. Much of what goes on today
in mathematics education is based on opinions that are so firmly held

that the thought of doubting them crosses very few minds. Yet most of these opinions have no empirical substantiation, and in fact many of them are, if not wrong, at least in need of serious qualifications. Examples are scattered throughout this monograph. These erroneous opinions are often the cause of inefficiencies in our educational programs. Until we learn to recognize them and until we start to pay more attention to facts than to opinions, no matter how plausible the latter may be, we can do nothing to eliminate these inefficiencies.

Thus, for example, in the late fifties and early sixties, some radical changes in education were suggested and, to some extent, put into effect. Examples are: teaching by film or TV, programmed instruction, team teaching, discovery teaching, individualized instruction. Proponents of these innovations provided persuasive arguments and made it seem very plausible that each would lead to substantial improvements in education in general and in mathematics education in particular. But the plausibility and validity turned out to be uncorrelated, and most of these highly touted "improvements" turned out not to be improvements at all, once the facts were in. Much time and money could have been saved if empirical studies of the validity of these suggested innovations had been carried out before widespread implementation was attempted.

Unfortunately, many of those most directly involved in mathematics education are still unaware of the empirical findings with respect to those particular innovations or, if they are aware, are somehow unwilling to acknowledge them. And there are many other aspects of our educational programs which are less effective today than the known facts suggest they might be.

In this survey, then, we will not concentrate on just these few innovations listed above, but rather will try to indicate how much is known about all aspects of mathematics education.

It needs to be emphasized that only mathematics learning is dealt with here. Of course students learn many other things, and the effects of pedagogical variables on student learning may, and in fact often do, vary from one subject matter to another. We must not generalize from what we know about the learning of mathematics to the learning of other things, such as reading or social studies for example, and vice versa.

Note that I used above the phrase "how much is known" rather than the phrase "what is known." It would require a very large effort

to exhibit in detail all that is known about mathematics education. I
have had neither the time nor the strength to carry out such a task
in full.

However, I believe that I have compiled a substantially complete
list of all the variables which have been studied for their effects on
mathematics education. In addition, I provide fairly accurate information on the amount of study which has been devoted in recent times to
each of these variables. This information was obtained by means of a
thorough search, for the period 1960-1976, through all the sources
listed in the Bibliographic Note which follows.

To record this information I have used the following format. I
divided the variables into five disjoint sets according to their loci.
I devote a separate chapter to each set. Thus Chapter Three discusses
those variables that are located within teachers, Chapter Four those
variables that are located within the curriculum, and so on.

The variables within each set were sorted out into relatively independent topics, and in some cases these last were even further subdivided into subtopics. I was unable to perceive any obvious taxonomy
for these topics, so they are presented, within each chapter, alpha-
betically by topic title.

For each topic, I provide a brief description of the variable (or
variables) and also indicate the kinds of experiments or other empirical
studies which have been used to investigate them. Whenever there have
been reviews of the empirical literature, I refer to them, and in some
cases I summarize some of the reviews. I also list any helpful bibliographies I have come across.

Whenever I have formed an opinion, on the basis of a more or less
careful study of the relevant documents, about the direction of the
findings with respect to a particular variable or about the need for
further research, I have felt free to express it.

An integer enclosed in square brackets appears after the title of
each topic or subtopic. This represents the number of references I have
found to empirical reports on that topic. The exact values of these
numbers are not important, but together they provide an indication of

the relative amount of attention which has been paid to different variables and may be of interest for this reason.

At the end of each of these five chapters is a Bibliography of the reviews, bibliographies, and other reports mentioned in the chapter. In addition there is a second bibliography of Illustrative Research Reports. Some readers of this survey may not be familiar with the nature of research in mathematics education. The references in the second bibliography are to articles, in journals which should be easy to find in any university library, which together illustrate the <u>kinds</u> of research studies which have been carried out. I wish to emphasize, however, that these articles are far too few in number to be able to illustrate the results of such studies. Nonetheless, these articles do provide enough examples of research in mathematics education to clearly demonstrate how different such research is from research in, say, mathematics itself.

Chapters Eight and Nine have exactly the same format as Chapters Three through Seven, but their contents were chosen differently. Tests are used in many different ways in mathematics education, and Chapter Eight surveys a variety of empirical studies of various aspects of mathematics tests.

One of the major objectives of mathematics education is to increase the problem-solving ability of our students, and Chapter Nine is concerned with empirical studies of various aspects of mathematical problem solving.

Chapters One and Two are introductory and provide what I hope will be helpful background information for the reader.

The final chapter contains my suggestions as to priorities for more extensive reviews of the available factual information on various aspects of mathematics education and for further research. I also append a few general comments.

In this survey I have emphasized empirical and factual information about mathematics education and have, implicitly at least, downgraded both expert opinion and common sense. This was done deliberately. It takes no more than a cursory look at the literature to make it clear that common sense and expert opinion have very little to contribute at

this time. This is particularly hard for mathematicians to understand
since they usually feel that the intellectual skills that are important
in doing mathematics ought to be equally important in discussing mathematics education. However, the field of mathematics education, in its
present state, is quite unlike that of mathematics and resembles more
closely the state which agriculture was in, in this country, several
generations ago. We have no established theory to provide a basis for
our discussions. Deductive reasoning now plays no useful role in mathematics education since educators have, in general, nothing to deduce
from.

My ambition in this survey was to make it clear that a large amount
of factual information could be made available in a reasonable length
of time by exploiting what is already known and by carrying out the
experimental studies which these known results indicate to be needed.
On such a foundation, theory building and mathematical reasoning could
be usefully carried out. Without such a foundation, we can only continue to spin our intellectual wheels.

There are four major sources of empirical information about mathematical education. First come the standard journals. The following list consists of journals which have provided me with useful information often enough in the past so that I find it worthwhile to scan all of them regularly.

Alberta Journal of Educational Research
American Educational Research Journal
American Mathematical Monthly
Arithmetic Teacher
Audio-Visual Communication Review
British Journal of Educational Psychology
British Journal of Educational Technology
California Journal of Educational Research
Educational and Psychological Measurement
Educational Research
Educational Studies in Mathematics
Florida Journal of Educational Research
Harvard Educational Review
Journal for Research in Mathematics Education
Journal of Educational Measurement
Journal of Educational Psychology
Journal of Educational Research
Journal of Experimental Education
Journal of Genetic Psychology
Journal of Research and Development in Education
Journal of Research in Science Teaching
Journal of School Psychology
Mathematics Teacher
Monographs of the Society for Research in Child Development
Peabody Journal of Education

Psychology in the Schools
Review of Educational Research
Saskatchewan Journal of Educational Research
Scandinavian Journal of Educational Research
School Science and Mathematics
Southern Journal of Education
Theory into Practice
Two-Year College Mathematics Journal

Next is the Educational Resource Information Center (ERIC). This operation, originally set up by the U. S. Office of Education and now supported by the National Institute of Education, has a number of different clearinghouses, each devoted to a particular aspect of education (one of them to mathematics, science, and environmental education) and each of which tries to obtain as complete a collection as possible of the ephemeral documents in its areas. Examples would be reports to U.S.O.E. or N.I.E., reports of evaluation teams to local school boards, copies of papers delivered at annual meetings of national educational organizations, etc.

Each document absorbed into the ERIC system is given an identifying number (e.g., ED 102 345) and an abstract is written. An annual index of these abstracts, with both subject and author indexes, is published.

Copies of most ERIC documents can be obtained either as photocopies or in microfiche form. For further information on the nature and operation of the ERIC system and for the location of depositories of ERIC microfiche write to:

ERIC Processing Reference Facility
Operation Research, Inc.
Information Systems Division
4833 Rugby Avenue
Suite 303
Bethesda, Maryland 20014

A third source of information on mathematics education consists of doctoral dissertations, several hundred of which are completed each year. Only a few of these, however, are published even in summary form. Extracting information from these dissertations is, unfortunately, not as straightforward as it is in the case of the documents listed under the first two headings above.

It is not difficult to find out what dissertations in mathematics education were completed each year (at least since 1960) and to read a brief abstract of each one. To go any further is a longer and more expensive task.

Almost all U. S. universities submit, to an organization called University Microfilms, a copy of each dissertation completed by their students. In addition to the dissertation, a brief abstract, no more than 600 words, prepared by the author of the dissertation is also submitted.

These abstracts are printed verbatim in a serial publication entitled *Dissertation Abstracts International* (formerly *Dissertation Abstracts*). An annual author index is also published. Most university libraries have these two publications.

The journal *School Science and Mathematics* has published each year since 1961 a list of the doctoral dissertations related to mathematics education (and science education) completed during the previous year. Since 1971, the *Journal for Research in Mathematics Education* has published a similar but annotated list of such doctoral dissertations.

Thus it is relatively easy to find out what doctoral dissertations have been completed and, for each one, to obtain some indication, through the author-prepared abstract, of the contents of the dissertation. But this is usually not enough. In order to decide whether or not the empirical information provided by a particular dissertation does have a bearing on a particular issue in mathematics education, it is almost invariably necessary to inspect the full dissertation.

Fortunately this is possible. Few libraries (I know of none) have substantial holdings of dissertations, but photocopies or microfilms of any of the dissertations abstracted in *Dissertation Abstracts International* can be obtained from University Microfilms. Of course, there is an appropriate charge for these.

A fourth major source of empirical information about mathematics education is the National Longitudinal Study of Mathematical Abilities (NLSMA). This was a five-year longitudinal study, lasting from September 1962 to June 1967, which was conducted by the School Mathematics Study Group. Over 100,000 students were involved in this study. They were in grades four, seven, and ten at the beginning of the study. These students were administered a battery of mathematical and psychological tests each fall and each spring of the study (except for those who started as tenth graders, who were only tested for three years). In addition, a good deal of information about the teachers, schools, and communities involved in the study was collected.

The study was a naturalistic one in that the choice of students to be included in the study and the choice of their teachers and textbooks was entirely in the hands of the cooperating schools. The only role of the School Mathematics Study Group was to collect data and, of course, at the conclusion of the study to carry out some analyses of the data.

A total of 32 reports on this study have been issued and are listed below. They have all been inserted into the ERIC system and hence are easily available.

The first three chapters of NLSMA Report No. 7 provide a good orientation to the purposes and organization of this study.

ED 044 277   No. 1   *X-Population Test Batteries*
Edited by James W. Wilson, Leonard S. Cahen,
Edward G. Begle.

ED 044 278   No. 2   *Y-Population Test Batteries*
Edited by James W. Wilson, Leonard S. Cahen,
Edward G. Begle.

ED 044 279   No. 3   *Z-Population Test Batteries*
Edited by James W. Wilson, Leonard S. Cahen,
Edward G. Begle.

ED 044 280   No. 4   *Description and Statistical Properties of X-Population Scales*
Edited by James W. Wilson, Leonard S. Cahen,
Edward G. Begle

ED 044 310   No. 5   *Description and Statistical Properties of Y-Population Scales*
Edited by James W. Wilson, Leonard S. Cahen,
Edward G. Begle

ED 044 281   No. 6   *Description and Statistical Properties of Z-Population Scales*
Edited by James W. Wilson, Leonard S. Cahen,
Edward G. Begle

ED 084 112   No. 7   *The Development of Tests*
Edited by James W. Wilson, Leonard S. Cahen,
Edward G. Begle
Authors, Thomas A. Romberg and James W. Wilson

ED 084 113   No. 8   *Statistical Procedures and Computer Programs*
Edited by James W. Wilson, Leonard S. Cahen,
Edward G. Begle

ED 044 282   No. 9   *Non-Test Data*
Edited by James W. Wilson, Leonard S. Cahen,
Edward G. Begle

ED 044 283   No. 10   *Patterns of Mathematics Achievement in Grades 4, 5, and 6: X-Population*
Edited by James W. Wilson, Leonard S. Cahen,
Edward G. Begle
Authors, L. Ray Carry and J. Fred Weaver

ED 045 447   No. 11   *Patterns of Mathematics Achievement in Grades 7 and 8: X-Population*
Edited by James W. Wilson, Leonard S. Cahen,
Edward G. Begle
Author, L. Ray Carry

ED 084 114   No. 12   *Patterns of Mathematics Achievement in Grades 7 and 8: Y-Population*
Edited by James W. Wilson, Leonard S. Cahen,
Edward G. Begle
Authors, Gordon K. McLeod and Jeremy Kilpatrick

ED 084 115   No. 13   *Patterns of Mathematics Achievement in Grade 9:*
Y-Population
Edited by James W. Wilson, Leonard S. Cahen,
Edward G. Begle
Authors, Jeremy Kilpatrick and Gordon K. McLeod

ED 084 116   No. 14   *Patterns of Mathematics Achievement in Grade 10:*
Y-Population
Edited by James W. Wilson, Leonard S. Cahen,
Edward G. Begle
Authors, Gordon K. McLeod and Jeremy Kilpatrick

ED 084 117   No. 15   *Patterns of Mathematics Achievement in Grade 11:*
Y-Population
Edited by James W. Wilson, Leonard S. Cahen,
Edward G. Begle
Authors, Jeremy Kilpatrick and Gordon K. McLeod

ED 084 118   No. 16   *Patterns of Mathematics Achievement in Grade 10:*
Z-Population
Edited by James W. Wilson and Edward G. Begle
Author, James W. Wilson

ED 084 119   No. 17   *Patterns of Mathematics Achievement in Grade 11:*
Z-Population
Edited by James W. Wilson, Leonard S. Cahen,
Edward G. Begle
Author, James W. Wilson

ED 084 120   No. 18   *Patterns of Mathematics Achievement in Grade 12:*
Z-Population
Edited by James W. Wilson and Edward G. Begle
Authors, Thomas A. Romberg and James W. Wilson

ED 084 121   No. 19   *The Non-Intellective Correlates of Over- and Under-*
Achievement in Grades 4 and 6
Edited by James W. Wilson and Edward G. Begle
Author, Kenneth J. Travers

ED 084 122   No. 20   *Correlates of Attitudes Toward Mathematics*
Edited by James W. Wilson and Edward G. Begle
Author, F. Joe Crosswhite

ED 084 123   No. 21   *Correlates of Mathematics Achievement: Attitude*
and Role Variables
Edited by James W. Wilson and Edward G. Begle

ED 084 124   No. 22   *Correlates of Mathematics Achievement: Cognitive*
Variables
Edited by James W. Wilson and Edward G. Begle

ED 084 125   No. 23   *Correlates of Mathematics Achievement: Teacher*
Background and Opinion Variables
Edited by James W. Wilson and Edward G. Begle

ED 084 126   No. 24   *Correlates of Mathematics Achievement: School*
Community and Demographic Variables
Edited by James W. Wilson and Edward G. Begle

1

The Nature of Mathematics and of Mathematical Objects

This survey is devoted to discussions of the effects of various educational variables on the learning of mathematics. These first two introductory chapters are designed to provide a framework for these discussions, a point of view that, I hope, will provide a focus and a way of unifying the very wide range of variables that have been investigated and the disparate methods that have been used to investigate them.

Mathematical Systems

One prerequisite to a study of the learning of mathematics is a clear understanding of the nature of the mathematics to be learned. I have found the following way of looking at mathematics a useful one from the point of view of mathematics education. I consider mathematics to be a set of *interrelated, abstract, symbolic systems*. In order to make clear the meaning of the preceding sentence I will try to explicate each of the italicized words.

Let us first consider, as an example, the part of mathematics dealing with the whole numbers. The basic symbols used are:

0, 1, 2, 3, 4, 5, 6, 7, 8, 9, +, -, ×, ÷, = , <, and >.

A *system* is based on these, the elements of the system being certain combinations or "strings" of these basic symbols. There are familiar grammatical rules which tell us that the following are grammatically correct phrases and hence elements of the system:

207, 121 + 2, 12 × 12 + 6 × 24;

while the following are not:

2 +, 12 × 24 ÷ , 2   7.

Also, there are other familiar grammatical rules which tell us that the following are grammatically correct sentences:

2 + 3 = 5, 2 × 3 = 6, 181 × 92 > 92 ÷ 181;

while these are not:

2 × 3 =, 2 + 3 = 6 +, 181 × 92 < .

Two operations, addition and multiplication, are defined next. The addition and multiplication tables:

+	0	1	2	. . .	9
0	0	1	2	. . .	9
1	1	2	3	. . .	10
2	2	3	4	. . .	11
.	.	.	.		.
.	.	.	.		.
.	.	.	.		.
9	9	10	11	. . .	18

×	0	1	2	. . .	9
0	0	0	0	. . .	0
1	0	1	2	. . .	9
2	0	2	4	. . .	18
.	.	.	.		.
.	.	.	.		.
.	.	.	.		.
9	0	9	18	. . .	81

are used to define sums and products of single digit elements of the system. For other elements, the familiar addition and multiplication algorithms are brought in.

The operations of subtraction and division are defined in terms of addition and multiplication, respectively, and algorithms for these operations are provided.

An important aspect of the structure of this system is that there are some laws or principles:

(a) $N + M = M + N$
(b) $N \times M = M \times N$
(c) $(N + M) + P = N + (M + P)$
(d) $(N \times M) \times P = N \times (M \times P)$
(e) $N \times (M + P) = (N \times M) + (N \times P)$
(f) $N + 0 = N$
(g) $N \times 1 = N$
(h) $N \times 0 = 0$

These are always true, no matter what whole numbers N, M, and P are.

Still another aspect of the system is the place-value principle, exemplified by

30 + 7 = 3 × 10 + 7.

By means of the algorithm and laws, the truth or falsity of any sentence involving these four operations and the equality sign can be determined.

A similar but somewhat more involved procedure can be used to determine the truth or falsity of sentences involving the symbols < and > . I need not go into the details of this.

The discussion so far is enough to demonstrate that we are dealing with an *abstract* (the symbols, so far, are meaningless), *symbolic* (we have nothing but symbols and strings of symbols) *system* (the principles above make it clear that we are dealing not with an arbitrary set of strings of symbols but rather one in which there are some regularities and some order).

It is possible for a student to learn much of the system of whole numbers, including the grammar of whole number expressions, the addition and multiplication facts, place value, the computational algorithms, and the structural principles, in a purely formal, rote fashion. Indeed, much of the teaching of arithmetic during the decades between the two World Wars was done in such a rote fashion, and the current cry of "Back to the Basics" seems to suggest a return to this practice.

However, we know, and this will be discussed at greater length in a later chapter, that a more effective way of teaching the whole number system is to *relate* it to another symbolic system which is less abstract and is closer to the real world. This is the system of all finite sets of concrete objects.

First we recall the operation of "pairing" the elements of two disjoint sets A and B, leading, in each case, to one and only one of the relations

$$A < B, \quad A \sim B, \quad A > B.$$

It is an experimental fact that, given two sets, the relation between them is independent of the way in which the pairing operation is carried out. Thus we can think of the pairing operation as an algorithm for ascertaining which of the above relations holds between two sets.

A number of principles concerning these relations are also experimental facts, such as:

when $A < B$ and $A' \sim A$, then $A' < B$, etc.

Two other operations on sets are also fundamental aspects of this system. The union operation associates with any two sets A and B, the set $A \cup B$ consisting of those objects, and only those, which belong to

3

A or to B or to both. The (cartesian) product operation associates
with a pair of sets A and B the set consisting of the union of disjoint
copies of B—one for each object in A. This is denoted by A × B.

Again, there are certain principles concerning these operations,
and again these are experimental facts. For example (all sets disjoint):

1) $A \cup B \sim B \cup A$,
2) when $A \sim A'$ and $B \sim B'$, then
    $A \cup B \sim A' \cup B'$,
3) $A \times B \sim B \times A$,
4) when $A \sim A'$ and $B \sim B'$, then
    $A \times B \sim A' \times B'$.

So far we have left implicit certain aspects of the concept "set"
which ought now to be mentioned. First, in the operation of pairing
the members of two sets, the sizes, weights, color, density, etc., of
the objects in the sets are ignored. Hence the relations $<$, $\sim$, and $>$
do not depend on these physical aspects of the objects constituting the
sets.

Second, when we have specified a particular set by indicating the
specific physical objects which it contains, we agree that it remains
the same set no matter how the physical location of the objects may be
changed.

For these reasons, we can consider this system to be an abstract,
symbolic one, even though it is, of course, much more concrete than is
the system of whole numbers.

It is clear that many sets are familiar, tangible things which are
frequently the objects of our attention in everyday life. Also, the
operations of pairing and forming the union (and to a lesser extent the
cartesian product) are frequently carried out by almost everyone (although the terminology may be unknown and the principles relating the
operations may be unrecognized).

Consequently, it is important to note that this system of finite
sets of concrete objects is closely related to the system of whole
numbers and that, in fact, the latter can be constructed from the former.
Let us review briefly how this is done.

We start by separating all sets into equivalence classes, where each set in any particular equivalence class is equivalent to every other set in that class and is equivalent to no set in any other class. We next attach to each equivalence class a number, i.e., an element of the symbolic system of whole numbers.

Now we can define the relations $<$, $=$, $>$ between whole numbers by referring back to the relations $<$, $\sim$, and $>$ between sets chosen from the equivalence classes corresponding to the numbers. Similarly, we can define the operation of addition of numbers N and M by choosing a set A with N elements and a (disjoint) set B with M elements and defining $N + M$ to be the number attached to the set $A \cup B$.

The operation of multiplication is defined in an analogous way using the cartesian product instead of the union.

When we look at the situation from this point of view, we see that the concepts of "number," "addition," "multiplication," "less than," "equals," etc., connect the symbolic system of whole numbers with the system of finite sets of concrete objects.

Thus the structure of mathematics has two parts. On the one hand, each mathematical system has its own internal structure. On the other hand, there are linkages, relations, between different systems which also contribute to the structure of mathematics.

It should be noted that we have considered, for illustrative purposes, only two mathematical systems: the whole numbers and finite sets of concrete objects. Of course there are many more mathematical systems, all linked together in one way or another. For example, before the end of elementary school the systems of one-, two-, and three-dimensional geometry are introduced, as well as the systems of non-negative rationals and the integers. In each case, the internal structure of the system is exposed to the students, but also the linkages between systems are not only made clear but are emphasized in the pedagogical procedures favored today.

A glance at the two mathematical systems sketched in the preceding paragraphs, or at any other mathematical systems, simple or complex, makes it clear that different *kinds* of mathematical ideas or, as I prefer to call them, mathematical objects are involved. Since the best way of teaching one kind of mathematical object may not be the best way of teaching another kind, it seems wise to recognize these differences explicitly. I have found the following four-way classification to be useful.

The four kinds of mathematical objects are:

1. *Facts*. That "three" is the word which is associated with the symbol "3" is a fact. That two plus three equals five is also a fact. Almost any mathematical system will contain some facts, particularly in connection with the notation used for the system.

There are two kinds of facts. The first example above is of an arbitrary fact. Presumably, this kind of fact is learned by rote, and the laws of rote learning which have been worked out by psychologists are applicable to the learning of this kind of mathematical fact.

Other mathematical facts, however, are not arbitrary since they can be deduced from other facts. For example, $7 \times 8 = 56$ is a fact, but it can be deduced from others:

$7 \times 8 = (7 \times 7) + (7 \times 1) = 49 + 7 = 56$. Similarly, the fact $2 + 3 = 5$ can be deduced from this diagram:

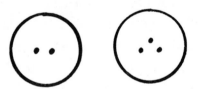

by remembering the definition of addition in terms of union of sets.

2. *Concepts*. As an example, consider the concept of "triangle." Triangles form a subset of the set of all geometric figures. In order to determine whether a given figure is a triangle, certain aspects, or "dimensions," of the figure need to be considered: the number of sides, the way the sides fit together, the straightness of the sides, etc.

At the same time, it is part of the concept of triangle that certain other dimensions are irrelevant, e.g., the lengths of the sides, the sizes of the angles, and the orientation of the figure.

Other examples of concepts are: three, rectangular array, mathematical sentence, fraction, congruence, greater than. (Note that in the case of the last three it is a relationship between pairs of objects that the concept specifies.)

Most (if not all) mathematical concepts, except for the most primitive undefined ones, such as set or element, are built up from earlier concepts by means of the usual logical connectives: *and, or, not, if-then, there exists,* etc.

3. *Operations.* An operation is a function which assigns mathematical objects to mathematical objects.

Examples of operations are: pairing the members of two sets, counting, forming the union of two sets, adding two numbers, measuring the length of a line segment, constructing the perpendicular bisector of a line segment, multiplying both sides of an equation by the same number, and representing addition of numbers by motions along a number line.

4. *Principles.* A principle is a relationship between two or more mathematical objects: facts, concepts, operations, or other principles. Any principle can be expressed as a mathematical theorem or axiom and every theorem or axiom expresses a principle (except for those that express a fact by stating the existence of a particular kind of mathematical object).

In terms of this discussion, we can think of mathematics as an immense three-dimensional linear graph, with facts and concepts as the nodes, and operations and principles as the connecting arcs. This point of view of mathematics as an intellectual structure is not opposed to that presented at the beginning of this chapter, but merely looks at mathematics from a slightly different perspective. Thus we can consider that part of the task of mathematics education is to assist our students to construct in their own minds selected parts of this mathematical network.

Of course there is more to it than this. We leave to the next chapter a discussion of the objectives of mathematics education, but almost everyone would agree that another important task is that of helping students learn how to apply mathematics to problems and situations arising in the real world. What little we know about the use of mathematics in problem solving is surveyed in Chapter Nine.

Note: The description of mathematics presented in this chapter is fairly austere and formal. Let no one conclude that I would ever suggest that mathematics be *taught* in such a fashion.

2

The Goals of Mathematics Education

What are the goals of mathematics education? What mathematics do
we want our students to learn? It is important to realize that there
are no clear-cut universally accepted answers to these questions. The
reason for this is simple. The number of kinds of individuals and organizations which have some input to the process of setting goals for
mathematics education is large and their interests are too varied to
make possible any unanimity.

Let us review briefly some of these sources of inputs to the goal-
setting process. First to come to mind, of course, are the classroom
teachers. Even if a set of goals is handed to the teacher by, say,
the school administration, without any chance for discussion between
the teacher and the goal setters, the teacher is nevertheless able,
within limits, to modify and to change the emphasis of the goals within
the classroom. But such lack of discussion is rare, and whenever a
committee or working group is asked to spell out goals for mathematics
education, it is practically certain that classroom teachers will be
well represented.

Equally influential in setting goals and specifying curricula are
mathematics educators. They are always well represented, along with
classroom teachers, on state and national goal committees. A substantial number of mathematics textbooks, particularly for the secondary
school, are authored by mathematics educators, and textbooks are a
powerful, if somewhat indirect, influence on mathematics education
goals. (See the appendix at the end of this chapter for substantiation

9

of this statement.) Probably a majority of the discussions of goals
which have appeared in the professional literature were authored by
mathematics educators, with classroom teachers running them a close
second. In his doctoral dissertation, R. N. Harding (1) provides
references to and a summary of the discussion of the last 50 years.
The dissertation of H. I. Oakes (2) is also relevant.

Students, on the other hand, seem to play little part in the setting of goals for mathematics education. There is little doubt that
they have fairly strong opinions about the specific mathematics topics
they are asked to learn, but they are not normally asked to participate
in goal discussions. There have, however, been some studies of situations in which students had some choice as to what they were to study.
We will discuss this matter in Chapter Seven under the heading Locus of
Control.

Often the local school system, with the assistance of some of its
teachers, will spell out in some detail the objectives of its mathematics program. This is also done at the county and state levels.
Illustrative examples are: Mesa, Arizona (3), Multnomah County, Oregon
(4), and Wisconsin (5).

State departments of education, in some of our states, can and do
exert an even stronger influence on mathematics education goals by
specifying what textbooks may be used in the public schools. State
legislatures can also exert influence. For example, in my own state,
California, a recent act of the legislature will make it mandatory for
a high school student to display a certain amount of proficiency in
mathematics (and of course in some other curriculum areas) in order to
receive a diploma.

As in the case of textbook choice by state education departments,
the various curriculum projects that flourished in the sixties provided
suggestions about the goals of mathematics education through the textbooks they constructed to illustrate their points of view about the
mathematics curriculum.

Another potent influence on the direction of mathematics education
rests in testing organizations. The lay public sets great store on
the tests produced by commercial test publishers. No innovation in the
mathematics curriculum can prosper unless it is recognized by the

commercial tests. While test publishers have tried to bring their
tests up to date and make them more in keeping with the recommendations
for change made in the sixties, their tests still are, on the whole,
more of a hindrance than an aid to the improvement of mathematics
education.

On the other hand, the College Entrance Examination Board (CEEB)
can claim considerable credit for sparking the new curricula and the
new goals for mathematics education which became prominent in the sixties. Concern as to whether examinations it was setting were influenc-
ing the high school mathematics curricula in a satisfactory direction
led CEEB to appoint a commission to review carefully the goals of
high school mathematics programs for college capable students. The
report of this commission (6) not only sparked a large amount of
thoughtful discussion about these goals but also provided the initial
direction for the School Mathematics Study Group (SMSG), the largest
organization, and the only national one, devoted, during the sixties,
to improvement of the mathematics curriculum. SMSG in turn, through
its sample textbooks, provided a considerable input to those who
ultimately set the goals for mathematics education.

Some other committees that functioned during the sixties led to
much discussion but seem, so far, to have had only minor influence on
mathematics education goals. Most prominent among these was the
Cambridge Conference (7).

Another testing organization that seems destined to have a powerful influence on mathematics education is the National Assessment of
Educational Progress (NAEP). This assessment program is conducted by
the Education Commission of the States, a consortium of several of the
states. It tests students and young adults in several subject matter
areas, including mathematics. Tests are administered periodically,
at intervals of three to five years, in order to detect changes in the
effectiveness of our overall educational programs. Before developing
their mathematics tests, a detailed list of objectives for mathematics
education was drawn up. An account of these particular objectives,
and of the procedures followed in developing them, is found in the
report by Norris and Bowes (8). It is interesting to observe the large
role played by lay persons in the development of these objectives.

With the exception of NAEP, lay persons, at least in recent years, seem to have had little influence on our mathematics programs. Even parents, despite such generalized cries as "Back to the Basics," rarely became well enough acquainted with school mathematics programs to be able to exert any substantial influence on them or on their goals.

Our colleges and universities have a strong influence on the goals of mathematics education at the high school level. They expect that entering students who are interested in science or engineering or the social sciences will come equipped with certain mathematics skills and understanding, and our high schools consequently include among their goals the development of these skills and understanding.

A potentially important influence on our choice of mathematics education goals, but one to which little attention seems to be paid at present, is to be found in surveys of what mathematics is actually used in science, industry, business, and the trades. Such surveys have been carried out. The ones by Miller (9) and Rahmlow and Winchell (10) are illustrative. Few of those directly involved in setting goals seem to be aware of such surveys.

Last, but of course not least, among those with influence on the goals of mathematics education are the professional mathematicians. They are, as a matter of course, along with the mathematics educators, included on committees to draw up mathematics education objectives and are usually listened to with great respect. However, their suggestions and recommendations are not always accepted. The reason for this, I believe, is that mathematicians do not answer a question more fundamental, but less often asked, than those leading off this chapter in the same way as the members of the various other groups listed above.

This question is: "Why teach mathematics at all?" Answers to this question generally fall on a continuum ranging from utility at one end to culture at the other. At the extreme utility end are those who feel that the only excuse for teaching mathematics is that it is useful in personal affairs or in business or as a prerequisite for other studies and that some grasp of mathematics skills and understanding is essential for survival in our society. At the other end are those who believe that mathematics is one of the supreme intellectual creations of mankind and consequently is such an important part of our culture that all students should be introduced to it.

With the exception of some classroom teachers, most of those listed above incline more toward the utility end of the continuum. The average mathematician, however, is somewhat inclined toward the culture end. (A perhaps extreme but nevertheless illustrative example is the statement by Braunfeld et al. (11).) Harding (1) found that mathematics educators also ranked the importance of the objective of becoming adept at applying arithmetic to problems of business and personal finance lower than did parents, school administrators, etc.

This list is now long enough. It shows that the number of different kinds of interests which insist on being involved in setting goals for mathematics education is so large that unanimous agreement on any set of clearly stated objectives is not to be expected.

This of course creates a dilemma. If we cannot agree on objectives, how can we decide what tests and other devices to use to evaluate student progress toward our objectives or to evaluate the effectiveness of new curriculum materials or pedagogical procedures? The usual response is to insist that each test be broad and wide-ranging enough to include a sufficient sample of the objectives of each interested party. It turns out that this is indeed a feasible procedure and in fact it was used in each of the NLSMA test batteries.

Recent Developments

Since 1960, there have been three major developments with respect to the goals of mathematics education. The first of these was the widespread acceptance of the fact that the outcomes of mathematics education (or of education in any subject matter area) can be multivariate. It is of course true that standardized tests have long provided more than one score, usually a computation score and a problem-solving score. In actual practice the two scores are usually combined, and each student is assigned a single number as a measure of his mathematics achievement.

Credit for calling attention to the fact that educational outcomes may be at different cognitive levels and that these levels form a taxonomy goes to Bloom and his colleagues (12). Various modifications of their taxonomy have been suggested. The one found most useful by NLSMA is described by the following matrix.

	Number Systems	Geometry	Algebra
Computation			
Comprehension			
Application			
Analysis			

The columns denote the mathematical content that might be considered and the rows denote four different cognitive levels which students might use in dealing with a specific content item. For the various cells in this matrix, NLSMA constructed sets of test items each of which was homogeneous with respect to both mathematics content and cognitive level required to work the items.

Brief descriptions of the content and the levels are:

Categories of Mathematics Content

> *Number Systems* – Items concerned with the nature and properties of whole numbers, integers, rational numbers, real numbers, and complex numbers; the techniques and properties of arithmetic operations.

Geometry - Items concerned with linear and angular measurement, area, and volume; points, lines, and planes; polygons and circles; solids; congruence and similarity; constructions; graphs and coordinate geometry; formal proofs; spatial visualization.

Algebra - Items concerned with open sentences; algebraic expressions; factoring; solution of equations and inequalities; systems of equations; algebraic and transcendental functions; graphing of functions and solution sets; theory of equations; trigonometry.

Levels of Behavior

Computation - Items designed to require straightforward manipulation of problem elements according to rules the subjects presumably have learned. Emphasis is upon performing operations and not upon deciding which operations are appropriate.

Comprehension - Items designed to require either recall of concepts and generalizations or transformation of problem elements from one mode to another. Emphasis is upon demonstrating understanding of concepts and their relationship and not upon using concepts to produce a solution.

Application - Items designed to require (1) recall of relevant knowledge, (2) selection of appropriate operations, and (3) performance of the operations. Items are of a routine nature. They require the subject to use concepts in a specific context and in a way he has presumably practiced.

Analysis - Items designed to require a nonroutine application of concepts.

For any particular test or test battery, the columns, but not the rows were refined to fit the particular group of students who were to take the test.

The background and evolution of this model is described in NLSMA Report No. 7. A discussion of similar models that have been used in other mathematics programs can be found in the chapters by Weaver and by Begle and Wilson in Begle (13).

Although this kind of model has not yet been widely adopted by the commercial test publishers, most of the research on mathematics education and most of the evaluations of curriculum materials and educational procedures now take cognitive levels into account.

A number of studies have been carried out, using a variety of subject matters, that together demonstrate that the six levels of the Bloom taxonomy are empirically as well as conceptually distinct. Studies of this kind, but restricted to mathematics, seem not to be known. One NLSMA Report, No. 27, however, is relevant. This study attacked the following question: Suppose we wish to predict how well a student is going to do on a particular mathematics scale (scale = a test restricted to one cell of the NLSMA matrix above) at the end of the school year. Which bits of information about this student which were available at the beginning of the school year, or even during the previous school year, would be helpful in predicting spring performance?

A stepwise regression procedure was used to investigate this question for each of the spring scales administered during the five years of the NLSMA project. A simple count of the results brings out two conclusions. First, the best predictors of spring computational ability were almost always measures of earlier computational achievement. Only rarely did a scale at a higher cognitive level contribute to prediction of a computational scale score. Second, the best predictors of the scales at the three higher levels were also somewhere in these levels. Very rarely did a computational scale contribute to the prediction of success on a scale at a higher cognitive level.

We thus have empirical evidence that computational achievement is something quite different from achievement at higher cognitive levels, but we do not have empirical information as yet which allows us to distinguish between these comprehension, application, and analysis levels as far as mathematics goes.

The second recent development was the realization and acceptance by all parts of the mathematics education profession that the broad goals of mathematics education do change with time and respond reasonably quickly to changes in our society. Consider two examples. In 1960, statistics was generally considered to be an upper level undergraduate course, and there were probably some statisticians who still felt that the entire subject should be postponed to graduate school. Today, it has been demonstrated that an honest introductory course in statistics is feasible at the high school level and that many of the prerequisite ideas about probability are feasible, interesting, and useful for junior high school students or even upper elementary school students. Many school systems have consequently added to their goals for mathematics education at the pre-college level that of providing students with some of the basic ideas of probability and statistics.

The mere fact of the feasibility of doing this was not what changed our goals. The change was undoubtedly due to the realization that much of the information that all of us receive from newspapers, magazines, and TV is statistical in nature and hence that it should be part of general education to learn something about interpreting statistical statements.

In 1960, the number of high speed electronic computers in existence was not very large, nor was the number of individuals able to make use of them. Today, the number of computers in existence has vastly increased, and their use by banks, retail stores, etc. means that they directly affect the life of the average citizen. In response to this change in our society, attempts were made, by the School Mathematics Study Group among others, to introduce some of the ideas about algorithmic data processing into the secondary school curriculum. A substantial course in computer oriented mathematics was found to be quite feasible for high school students. In addition, some introductory ideas, in particular those relating to flow charting, turned out to be not only perfectly feasible for junior high school students but also quite useful pedagogically in developing other mathematical ideas for these students. In fact, the role of computers in our society is so important that a national committee (14) has called for a course on "computer literacy" to be made a requirement for all secondary school students.

The third major development actually dates back over 25 years to the time when Ralph Tyler (15) pointed out that we should evaluate educational program only in terms of their effects on students, not on the characteristics of the programs. Since none of us is a mind reader, we are unable to get directly into the minds of our students to find out how much they have learned from a particular educational program, nor can we find out directly what effect the program has had on student attitudes, self-concepts, etc. We are forced, therefore, to resort to various probes, such as asking questions, posing problems to be solved, observing the students in mathematics-related situations in order to estimate the effects of the educational program on these students. In other words, we can only evaluate educational programs by observing changes in student behavior brought about by the program.

An objective stated in terms of student behavior is called a "behavioral objective." The following is a behavioral objective:

At the end of the (geometry) course, the student will be able to carry out and justify each of the constructions studied during the course.

The following is not a behavioral objective:

During the (geometry) course, the students will be exposed to a great variety of constructions.

This says nothing about the effects of the exposure on the students. Indeed, this objective could be satisfied without the students learning anything about constructions. Many of the stated objectives put forth in the past were more like the second of these than the first.

During the fifties and sixties little use was made of behavioral objectives, although the NCTM Secondary School Curriculum Committee (16) emphasized the need to state objectives in terms of student behavior.

At the beginning of this decade, however, the situation changed suddenly and drastically. Probably as a result of public unhappiness over the ever rising cost of education and the resulting demand from the public that schools be held accountable for the funds they received, school administrators decided that it was necessary to spell out in some detail exactly what they were trying to do. Behavioral objectives seemed ideal for this purpose, and so teams of teachers were called in to rewrite, in behavioral form, the objectives for each of the subjects taught in the school including, of course, mathematics.

The results were not very satisfactory. In many cases the objectives were refined and subdivided into so many tiny bits that it was hard
to see the woods for the trees. In some cases it was forgotten that the
converse of a true proposition need not be true, and some objectives
were included that were indeed in behavioral form but that had nothing
to do with mathematics. A major problem was that very few of the objectives were above the cognitive level of computation.

When mathematicians looked at these lists they immediately jumped
to the conclusion not that the quality of some of the lists was low
(which was true) but rather than behavioral objectives could be written
only at the lowest cognitive level (which was false). Evidently these
mathematicians were unaware of the fact that NLSMA had long since demonstrated that it was possible to write test items at higher cognitive
levels and hence that it was perfectly feasible to state behavioral
objectives at these higher levels. They seemed equally unaware that they
had been using behavioral objectives implicitly all their professional
lives in the quizzes, tests, and examinations which they gave their
students.

The situation as it presently stands is unsatisfactory. Classroom
teachers find it difficult to write objectives at the higher cognitive
levels. Mathematicians have shown themselves to be too naive about
objectives to be of much help. An increased use of qualified mathematics
educators might lead to improvement.

Current Concerns

There are three movements going on in our society which might affect
the objectives of mathematics education. The first of these has to do
with measurement. The United States is the last major country still
using the English system of measurement. That system will probably be
replaced here by the metric system in the not too distant future. This
raises two questions. When the metric system becomes official in this
country, can we forget all about the English system and, in particular,
will we be excused from converting measurements from one system to the
other? Will it be reasonable to reduce, in our curriculum, the emphasis
on common fractions?

My guess is that the answer to both of these questions will be: "Probably not." While the evidence from other countries which have already switched from the English to the metric system indicates that it will take only a few years for the average adult to become accustomed to the new system, the need to convert from one system to the other will linger on for awhile. Furthermore, unless we wish to ignore history, there will be a perhaps small but nevertheless continuing need to translate measurements made in the past into the metric system once we have accommodated ourselves to it.

In any case, conversion is now much easier than it used to be. To go from feet to meters, for example, is conceptually no different from going from feet to yards. Only the arithmetic is more difficult. But with hand-held calculators now so widely available the arithmetic is no longer a problem.

In short, it seems likely that the English system of measurement will fade out slowly rather than instantaneously, and hence that the part of our mathematics education objectives that have to do with measurement can be modified slowly, with time for thought and experimentation, rather than all at once.

As for fractions, there will not be much change. While the English system of measurement will be replaced by the metric system as a result of legislation, no amount of legislation will cause Nature or our culture to present us only with fractions whose denominators are powers of ten. The day will continue to be one-seventh of a week, the aces will continue to constitute one-thirteenth of a deck of cards, and the chance of rolling a seven with a pair of honest dice will still be one-sixth.

Insofar as we are faced with problems which go beyond the measurement of lengths and weights, the metric system will not provide us with simpler fractions than the ones we have to work with now. Consequently, little change in our objectives with respect to rational numbers is called for.

The second movement is the technological one which has reduced the cost of the just mentioned hand-held calculators so that they are now within the price range of almost every student. These calculators can carry out calculations with whole numbers and with decimals more rapidly and more accurately than can most eighth grade students. They also make computation with fractions a lot less burdensome.

The easy availability of these calculators poses a question. Should we assume that they are and always will be available to our students, and if so should we require less in the way of computational skills, or should we keep our objectives unchanged and use calculators only as pedagogical devices to help students improve their numerical skills and understanding?

It is to be hoped that a final answer to this question will be postponed till after a good deal of experimentation with and evaluation of various ways of incorporating calculators into our mathematics education programs. In any case, whatever the answer is, it will have crucial implications for our objectives for mathematics education.

Finally, a number of states are taking, or contemplating taking, actions similar to the California law on high school diploma requirements mentioned earlier. Plans are to supplement or even replace the usual graduation requirements, stated in terms of courses completed satisfactorily, with examinations, not necessarily in the usual paper-and-pencil format, of student competence in selected areas. Different states are preparing various versions of this overall plan. For a useful review of the current state of affairs see Spady (17).

This movement should be watched carefully. It offers some important advantages, but there are also some potential difficulties. On the positive side, the needed tests can only be constructed on the basis of a clear and specific set of objectives for mathematics education. An opportunity is thus provided to call on a wider range of concerned indi-viduals to state these objectives and to revise and bring up to date sets of objectives now in use but perhaps outmoded in view of the happenings of the last decade or two.

A potential danger lies not so much in the fact that "competencies" will likely be interpreted to mean indicators of successful performance in real-world activities, for a greater emphasis in this direction seems needed, but rather in the fact that it will be too easy to forget that tomorrow's real world will probably be quite different from today's.

In the twenties and thirties in this country, the pre-college mathematics curriculum was considerably watered down under the influence of some educators who argued that students should not be asked to learn mathematical facts and skills that were not used by the average citizen in everyday life. It took World War II to remind educators that the world our students will have to cope with after they finish school may, and probably will, be more complicated and demanding of quantitative skills than the world outside their present-day school windows. It would be disastrous to repeat the mistake of watering down the curriculum.

Another possible danger lies in the possibility that the minimal competencies the stated objectives spell out will become maximal and that teachers will concentrate so hard on getting all their students to reach these objectives that they will have no time to consider further objectives for those students who are capable. Closely connected is the fact that the minimal objectives for some students should be more extensive than for others. For those students who are interested in and capable of a career involving social, physical, biological, or engineering science, the mathematical prerequisites should be part of their high school objectives, even if these prerequisites have no direct application in today's real world.

Those potential dangers, however, particularly if they are kept in mind, are probably outweighed by the advantages of moving from the indirect measure of student achievement, number of courses completed, to the more direct measure of student competency as revealed by suitable tests.

One of the first decisions by the School Mathematics Study Group was to prepare sample textbooks to illustrate what SMSG meant by an emphasis on concepts and understanding. It was felt that an outline or syllabus would not be sufficient. However, a number of mathematics educators told us that this was wishful thinking, that classroom teachers normally ignored the textbooks given them and taught what they wanted and how they wanted.

As it turned out, these mathematics educators were mistaken. NLSMA provided considerable evidence that mathematical topics and mathematical emphases included in a textbook do get through to the students. As an illustration, consider scales X509, Multiplication of Fractions; X511, Division of Fractions; X523, Whole Number Structure; and X525, Algorithms, described in NLSMA Report No. 4, which were administered at the end of grade 6 to those students who were entering grade 4 at the beginning of NLSMA. The first two of these were at the cognitive level of computation. The other two were at the cognitive level of comprehension. X523 measured understanding of properties of whole numbers, and X525 measured understanding of the rationales of computational algorithms.

Three conventional textbook series for grades four through six, in addition to the SMSG series, were included in the NLSMA analysis. The conventional textbooks emphasized computation more than the SMSG textbooks did, while the latter emphasized understanding and concepts more than the conventional textbooks did. As reported in NLSMA Report No. 10, the students using the three conventional series scored higher than the SMSG students on scales X509 and X511, while the SMSG students scored higher than the conventional students on X525 and much higher on X523. Differences in emphasis between these textbook series were clearly reflected in differences in student achievement.

At higher grade levels the contrasts, though not as large, were equally clear. Secondary level, as well as elementary level, textbooks do have an effect on student learning and are evidently not ignored by teachers.

Bibliography

1. Harding, R. N. The Objectives of Mathematics Education in Secondary Schools as Perceived by Various Concerned Groups. Doctoral dissertation, University of Nebraska, 1968.

2. Oakes, H. I. Objectives of Mathematics Education in the United States from 1920 to 1960. Doctoral dissertation, Columbia University, 1965.

3. Mesa Public Schools, Arizona. Mathematics Education: Student Terminal Goals, Program Goals, and Behavioral Objectives. ED 086 560.

4. Multnomah County Intermediate Education District, Portland, Oregon. Course Goals in Mathematics, Grades K-12. Critique Draft. ED 073 553.

5. Chandler, A. M., and others. Guidelines to Mathematics, K-6. Key Content Objectives, Student Behavioral Objectives, and Other Topics Related to Elementary School Mathematics. ED 051 185.

6. College Entrance Examination Board. *Program for College Preparatory Mathematics*. Report of the Commission on Mathematics. New York: CEEB, 1959.

7. Cambridge Conference on School Mathematics. *Goals for School Mathematics*. Boston: Houghton Mifflin Company, 1963.

8. Norris, E. L., and Bowes, J. E. National Assessment of Educational Progress, Mathematics Objectives. National Assessment Office, Ann Arbor, Michigan. ED 063 140.

9. Miller, G. H. A National Study of Mathematics Requirements for Scientists and Engineers. Final Report. ED 022 679.

10. Rahmlow, H. F., and Winchell, L. Mathematics Clusters in Selected Areas of Vocational Education, Report Number 8. Washington State University, Pullman. ED 010 659.

11. Braunfeld, P., Kaufman, B. A., and Haag, V. Mathematics Education: A Humanist Viewpoint. *Educational Technology*, November 1973, pp. 43-49.

12. Bloom, B. S. (Ed.). *Taxonomy of Educational Objectives*. New York: David McKay Company, Inc., 1956.

13. Begle, E. G., (Ed.). *Mathematics Education*. Sixty-ninth Yearbook of the National Society for the Study of Education, Part 1. Chicago: University of Chicago Press, 1970.

14. Conference Board of the Mathematical Sciences, Committee on Computer Education. *Recommendations Regarding Computers in High School Education.* April 1972. ED 064 136.

15. Tyler, R. W. *Basic Principles of Curriculum and Instruction.* Chicago: University of Chicago Press, 1950.

16. National Council of Teachers of Mathematics. The Secondary Mathematics Curriculum. *Mathematics Teacher,* Vol. 52 (1959) pp. 389–417.

17. Spady, W. G. Competency Based Instruction: A Bandwagon in Search of a Definition. *Educational Researcher,* Vol. 6 (1977) pp. 9–14.

3

Teachers

There is no doubt that teachers play an important part in the learning of mathematics by their students. However, the specific ways in which teachers' understanding, attitudes, and characteristics affect their students are not widely understood. In fact, there are widespread misconceptions, on the part of not only lay persons but also mathematics educators, about the ways in which teachers influence mathematics learning by their students.

Teachers have been looked at from a number of different points of view. However, the focus has not always been on the effect of teachers on student learning. One of the largest investigations of teachers, reported by Ryans (1), sought only to identify those teacher characteristics which most readily distinguished between experienced and inexperienced teachers. Three personality facets developed by this investigation for this purpose have since appeared in other studies and will be mentioned later.

Ryans was interested in teachers in general, not just in mathematics teachers. However, there have been a number of studies of characteristics of mathematics teachers. These seem to fall into two sets: studies of teacher attitudes and studies of teacher knowledge of mathematics.

Most, but not all, of these studies were concerned with elementary school teachers. Most of them seem to take it for granted that it is important that teachers have a positive attitude toward the subject matter they are teaching. As we shall see, such attitudes may not be as important as they are usually assumed to be. In any case, contrary to commonly accepted belief, elementary school teachers seem to have attitudes toward mathematics which are neutral at worst.

Knowledge of Mathematics [30]

Almost all of these studies investigate either pre-service or in-service elementary school teachers. Parallel to the case of the previous category, it seems to be taken for granted that it is important for a teacher to have a thorough understanding of the subject matter being taught. The question is never raised in these studies as to how thorough the understanding needs to be.

Studies of the effects on student learning of teacher attitudes and of teacher knowledge are discussed below. Until these effects are better understood, it is hard to see the necessity for further studies of these characteristics per se.

Effectiveness

While there has been a substantial amount of study of the characteristics of teachers in general, these studies are only a small part of an enormous literature, much of which is devoted to searches for those characteristics that identify the effective teacher. The importance of such searches cannot be over-estimated. School administrators charged with the hiring of new teachers or making decisions on the retention of probationary teachers need some way of distinguishing the effective from the ineffective teacher. Those responsible for the design and operation of teacher training programs have the same need.

Consequently, these searches have been going on for a very long time. Unfortunately, none of them have been successful. Despite all our efforts we still have no way of deciding, in advance, which teachers will be effective and which will not. Nor do we know which training programs will turn out effective teachers and which ones will not.

Before going on to a look at the relevant empirical literature, we need to comment on the meaning of "effective." If we agree that the focus of our educational enterprise is student learning (in the broad sense, including affective as well as cognitive learning), then we should base our definition of teacher effectiveness on the degree to which the teacher's students learn, or, at the very least, we should validate any other kind of definition that we might adopt against measures of student learning.

There are indeed practical difficulties connected with the measurement of student learning, especially affective learning or cognitive learning at higher cognitive levels. In fact, there are some individuals who find these difficulties too severe to be overcome and who therefore turn to definitions of teacher "effectiveness" in terms of teacher characteristics or teacher behavior which seems "reasonable," even though it has no demonstrable relationship to student learning. Such individuals remind me of the drunk who lost his keys in the middle of the block but insisted on searching for them at the corner because that was where the street light was.

There are, however, various ways in which student learning can be measured and used to determine teacher effectiveness. Before describing the particular procedure that today seems most appropriate, I need to emphasize that any measure of teacher effectiveness is relative. There is no such thing as the absolute effectiveness of a teacher. We can only deal with the effectiveness of a teacher relative to the effects of a pool of teachers on the learning of their students. This relativity of the measure of teacher effectiveness, fortunately, is not a serious matter, since in most cases when we wish to apply the notion there is a well-defined pool. Thus, for example, if a school administrator wishes to hire a new mathematics teacher and wishes to select on the basis of effectiveness, the pool is well-defined--it is the set of all potential applicants for the position.

As a first approach, we might compute, for each teacher in the pool, the average score on some appropriate test of all of that teacher's students at the end of a teaching episode and subtract from that figure the average score on the same test at the beginning of the teaching episode. The difference would be an indication of the teacher's effectiveness, and we could use these difference scores to rank order the teachers. However, this procedure has a number of defects. One of them is technical and has to do with the lack of reliability of difference scores. Another, more serious, is that this computational procedure does not take into account the possibility that one teacher's students might be brighter than those of another teacher. If so, the first teacher might receive an unfairly higher effectiveness score than the second teacher.

In order to get around these difficulties the standard procedure these days is to compute, for each student, a "predicted" score on the final test. This is done by a multiple regression analysis that uses the pre-test scores, as well as measures of student ability, as covariates. This "predicted" score can be thought of as the average score of all the students of all the pool of teachers having the same pre-test score and the same measures of ability.

The predicted score is then subtracted from the actual final score, for each student, to obtain a "residual" score. If this is positive, the student has performed above average, or above prediction. If negative, the student has done less well than predicted. The average of these residual scores, for all the students of a particular teacher, is then a measure of that teacher's effectiveness.

As pointed out earlier, this effectiveness measure depends on the other teachers in the pool. It also depends on the pool of students. It also depends on the particular test used, and it sometimes makes sense, as we shall see later in this chapter, to compute effectiveness scores with respect to different kinds of tests. In any case, whoever is computing the effectiveness score presumably has some specific goals, perhaps including some affective goals, in mind and chooses the tests accordingly.

Another problem in connection with the definition of teacher characteristics arises from the desire of those who are more concerned with long-term than with immediate teacher effects. Thus it is often stated that the real measure of a teacher is how well his students do in later courses or in real-world activities after schooling is finished. Such delayed measures of teacher effectiveness are, unfortunately, meaningless since they cannot take into account the various influences on the students between the end of the instructional period and the administration of the criterion test.

But the definition of teacher effectiveness, in terms of student learning, has a final and much more serious difficulty. It is ex post facto. The effectiveness of a teacher cannot be calculated until after the teacher has gone through at least one teaching episode.

What is needed, for example for school hiring purposes or for designing teacher preparation programs, are estimates of teacher effectiveness which can be obtained in advance. Accordingly, searches have been going on for over three quarters of a century for indirect measures of teacher effectiveness which, on the one hand, can be obtained without waiting and, on the other hand, are good predictors of the effectiveness of a teacher as measured by student learning, using the procedure discussed above. As already mentioned, these searches have had no success so far.

Thus, despite all these problems, the only satisfactory way we have of defining teacher effectiveness is in terms of student learning. The review of the literature presented below will provide a justification for this statement.

We will first review the results of these searches with respect to teachers in general before confining our attention to the specific case of mathematics teachers. Fortunately, we can be brief because three substantial reviews are already available. The first of these was done by Barr (2), who directed the dissertations of about 75 doctoral students, each of whom investigated some aspect of teacher effectiveness. We need mention only a few of the highlights of his report on these dissertation studies.

In the introductory chapters, Barr points out some of the difficulties in studying teacher effectiveness. For example, teachers may have a certain amount of freedom with respect to their educational goals, so the criteria can vary from teacher to teacher. Also, teachers are often called on to do more than teach. They may be expected to counsel students outside of class, to participate in community activities, etc., and these things are rarely taken into account in measuring teacher effectiveness.

Barr points out also that three different procedures were being used in the twenties, where his report starts, (and as a matter of fact are still being used) to measure teacher effectiveness. The first method relies on overall judgements or ratings of teachers made by administrative personnel, peers, and students. The second asks how well the teacher scores on traits, abilities, and skills and understanding that are presumed to be attributes of the effective teacher. Included here would be the teacher's grades in his formal teacher preparation program. The final method is the one discussed earlier that relies on student gains, both cognitive and affective.

Barr's fourth chapter is of considerable interest since he lists the data-gathering devices that were applied to teachers in one or more of the studies under review. Their variety is amazing. They range from tests of subject matter knowledge; through measures of various personality traits, of mental hygiene, of attitudes toward teaching, of physical fitness, of mental ability, of academic achievement, of citizenship, of culture, etc.; to ratings resulting from interviews and whatever information can be gleaned from audio recordings or autobiographies. No stone was left unturned.

Barr summarizes in Chapter VI seven studies in which factor analysis was used. In four of these studies, student achievement was measured and showed up on one factor. In the same studies, ratings assigned by supervisors of some sort were also recorded. These showed up, in each case, on a different factor. The other three studies did not measure student gains but did include ratings by different kinds of individuals or of different aspects of the teachers. More than one factor appeared in each study. Thus the first four studies indicate that ratings of teachers are not closely connected to student achievement, while the other three indicate that ratings provide a variety of bits of information about teachers rather than a single measure.

Chapter VII presents an interesting hypothesis, namely that there are a limited number of qualities each of which is essential for good teaching. If a teacher falls below the cutoff point on even one of these, superior performance on the others cannot make up for it. This hypothesis differs from the usual ones that take different qualities to be additive in their effects, so that a poor performance on one can be compensated for by superior performance on others. Unfortunately, there seems to have been little or no follow-up on this hypothesis.

Chapters VIII, IX, and X are devoted to personality variables and to motivation. Little in the way of positive findings is presented. The need for better and more precise definitions and measuring instruments is emphasized.

Chapter XI is well worth reading in full. It lists 35 assumptions which had been made by one or more of the investigators whose work was reviewed. Many of these assumptions were, and still are in many cases, considered by some to be self-evident truths which it would be nonsense to question.

In the final summary chapter, Barr observes that: "Teaching does not take place in a vacuum. . . . Effectiveness does not reside in the teacher per se but in the interrelationships among a number of vital aspects of a learning-teaching situation and a teacher."

This point--that the outcome of teaching does not depend just on the teacher but rather is the result of a complex interaction between the teacher, the students, the subject matter, the instructional materials available, the instructional procedure used, the school and community, and who knows what other variables--is one that will receive further support as we proceed with our review of the literature.

The second of the three reviews was carried out for the U. S. Air Force by Morsh and Wilder (3). Their coverage was wider than that of Barr, since they attempted to review all of the empirical studies which appeared during the first half of the century. Again we will mention only some of the high spots of their review.

The review is divided into two parts. The first part is concerned with criteria of teacher effectiveness. As did Barr, Morsh and Wilder observe that a variety of rating procedures had been used and that there was a lack of uniformity in these procedures across school systems and across teachers' colleges. They also note that administrative ratings of general effectiveness are generally consistent, but that ratings of more specific attributes of teachers are usually less reliable.

The reviewers noted that there had been a few studies using objective observations of teachers. They felt that such studies were still in their infancy but expressed optimism about this approach to the study of teacher effectiveness. We discuss this matter in Chapter Seven.

Morsh and Wilder's conclusions about the comparison of ratings of teacher effectiveness with measures based on student learning, which were based on a wider sample than that used by Barr, are quite similar to Barr's. This kind of indirect measure was not very useful as a measure of teacher effectiveness. Different kinds of raters agree among themselves but not with other kinds of raters, and their ratings do not correlate well with student learning.

The second part of the review reports on studies of a variety of teacher traits and qualities presumed to be related to teacher effectiveness. These traits and qualities ranged from intelligence and academic achievement through age, experience, participation in extracurricular activities, general culture test scores, SES, and interest in teaching to such indirect measures as voice characteristics and photographs. In each case, correlations with other measures of teacher effectiveness, while showing a wide range in general, were rather low.

The following summary statistics may be of interest: In the entire review, 687 correlations of pairs of measures of teacher effectiveness are reported. Of these, 179 used pupil achievement as the basis of one of the measures. Of these, 14 used achievement in mathematics. For the 179 cases in which pupil achievement was measured, the range of the correlations was from −0.69 to 0.81. The average was 0.065. For the 14 in which mathematics achievement was measured, the range was −0.52 to 0.45, with an average of 0.025.

Judging from these two reviews, attempts to measure the effects of of teacher characteristics on student achievement have had little success.

The third review was carried out by B. Rosenshine (4) and is devoted largely to studies conducted since 1960. The title of his book suggests a concern with teacher behavior rather than teacher attributes. However, some of the behavior he reports seems, to me at least, to be an indicator of persistent teacher traits which are probably not amenable to training, and I therefore think of them as characteristics of the teachers. Thus he devotes a chapter to reports on studies of teacher approval and disapproval, behavior which probably can be easily modified by training, but follows it with a chapter on teachers' cognitive behavior, much of which seems much less changeable.

Before going on, it is essential to note that Rosenshine reports only on studies which compare teacher behavior with student achievement and not, as was so often the case in the first two reviews, with other measures of teacher effectiveness.

It is also important, however, to notice that the literature on teacher behavior and teacher traits is fuzzy because investigators using the same name for a kind of behavior or a trait often use different measuring devices, and conversely, investigators using the same measuring device sometimes give different names to the same kind of behavior.

Much of Rosenshine's report uses the three teacher factors, mentioned above, which were developed by Ryans in (1). These are: X, warm, understanding vs. aloof, egocentric teacher behavior; Y, businesslike, systematic vs. unplanned, slipshod teacher behavior; and Z, imaginative, enthusiastic vs. dull, routine teacher behavior.

Rosenshine found sixteen studies in which a measure of teacher "warmth" was obtained. However, only two of these used Ryan's X rating scale, thus supporting the comment above about the use of the same name for different measures. Of these sixteen, four supplied significant evidence of the positive effect of the teacher warmth on student achievement, four others showed a mixture of positive and nonsignificant effects, and the remainder showed no significant effects at all.

Rosenshine found ten studies in which Ryan's measure of businesslike teacher behavior, or some similar measure, was obtained. Of these, seven showed a significant positive relationship with student achievement, one showed both positive and nonsignificant relationships, and one showed only nonsignificant relationships.

Another 16 studies were located in which teacher measures were obtained which seem to be closely related to Ryan's Y factor. Of these 16 studies, nine showed a positive relationship with student achievement, five showed both positive and nonsignificant relationships, and two showed no significant relationships.

Five studies were located that did not fit into the Ryan classification scheme. In these, a measure was obtained of students' perceptions of the difficulty of lessons. The results of comparing these with student achievement were mixed, but it is not clear whether this was due to the variable or to its measure.

Finally, along these lines, Rosenshine found six studies in which teacher enthusiasm was measured, although it is not clear how these compared with Ryan's Z factor. In each of them a positive relationship with student achievement was found, although in three there were also other nonsignificant findings.

In their reviews, Barr and Morsh and Wilder found that ratings of teachers did not have much relationship with student achievement. Rosenshine found 21 recent studies (all but four after 1960) which together show a little more promise. Of 12 of these which obtained ratings of teachers by their students, five showed significant relationships to student achievement, one had mixed results, and six had only nonsignificant results. Twelve studies which used general observers, or more than one kind of observer, were less impressive. Three showed significant relationships, two had mixed results, and seven showed only nonsignificant relationships. However, none of the correlations were negative.

Three studies used self-ratings. Two of these demonstrated significant positive relationships with student achievement, while the other provided only nonsignificant comparisons. On the other hand, two studies which involved peer ratings had only nonsignificant results.

In Rosenshine's final chapter some teacher background characteristics were extracted from studies of teacher behavior. In nine studies of the amount of teaching experience, all but two showed positive correlations with student achievement, but in only two cases were the comparisons statistically significant. In six studies of the amount of teacher professional preparation, only two significant relationships were found and both favored a *lesser* amount of preparation.

In six studies of teacher knowledge or aptitude, only two showed significant positive relationships. In another six studies of teacher attitudes, only two resulted in significant relationships with student achievement, and in one of these cases a negative attitude correlated with higher achievement.

In general, then, Rosenshine's review is not very encouraging. Ratings of teachers in the studies he reviewed, except perhaps for self-ratings, are no better correlated with student achievement than was the case for the studies in the Barr and the Morsh and Wilder reviews of older studies. Nevertheless, faith in ratings persists, as is made evident in a recent review of effective college teaching prepared by Kulik and McKeachie (5). A positive note does show up in the Rosenshine review in that two of the three Ryan factors seem to be quite strongly related to student achievement, and the third factor also leans in this direction, although not quite as strongly. But, as already stated, it is not clear whether the teacher qualities described by these three factors are amenable to training.

While these three reviews cover much of what is known about those teacher characteristics which affect student learning, they do not provide the complete story. Thus, for example, all three mention the amount of professional training of the teachers as an important characteristic. A recent study by Popham (6), however, casts serious doubts on the importance of professional training.

For those who wish to look further at studies of teacher effectiveness, not restricted to the case of mathematics teachers, four recent bibliographies are available, assembled by Baral (7), Blount (8), the Canadian Teachers' Federation (9), and Scott (10).

Finally, there is a serious threat to any of the positive findings in the three major reviews surveyed above. The very concept of the effectiveness of a teacher may not be valid. There is some indication that effectiveness is not a stable teacher characteristic, but it can vary over time. Rosenshine (11) has collected some information pointing in this direction.

Let us now turn our attention to studies of the effectiveness of mathematics teachers. What we are looking for are characteristics or attributes of teachers which correlate positively with, and hence predict to some extent, mathematics learning by their students. Several different kinds of characteristics have been studied.

A) Background Variables [30]

In the principal's office in any school there is a folder for each teacher in the school which contains standard background information about the teacher: college degrees earned, amount of professional preparation, number of years of teaching experience, sex, etc. In general, these variables, while they may be positively related to student learning, are rather weak predictors and contribute little to an explanation of teacher effectiveness.

Information with respect to such variables was collected for a very large number of teachers in the National Longitudinal Study of Mathematical Abilities, and it seems appropriate to summarize this information here.

NLSMA also collected information on a large number of student variables, in addition to scores on mathematics scales, and on many school variables. These non-mathematics variables were called "C-Variables" by the NLSMA staff since they were used to classify students and teachers into sets differing on these variables so that differences in mathematics achievement between these sets could provide information about the influence of these variables on the learning of mathematics.

The mathematical and psychological variables studied by NLSMA are described in NLSMA Reports Nos. 4, 5, and 6, and the individual test items are reproduced in NLSMA Reports Nos. 1, 2, and 3. Teacher and school and community measures are reproduced in NLSMA Report No. 9.

The C-Variable analyses are described in NLSMA Reports Nos. 21-25 and are summarized in NLSMA Report No. 26. For illustrative purposes we describe one such analysis: the one for which the C-Variable was the teacher variable Number of Years Teaching. The first step had been to select about 50 mathematics scales which covered the grade span from 4 to 11 and which sampled as far as possible all the cells of the matrix described in Chapter Two. Next, at each grade level four sets of students were selected. Since the NLSMA Advisory Panel had shown interest in possible sex differences, there were two sets of boys and two of girls, so the two sexes could be analyzed separately. For each sex, one set consisted of students who had used a conventional textbook during the grade in question. The other set had used an SMSG textbook, since this was felt to be representative of modern textbooks.

Each of these sets was then independently subdivided into three subsets, H, M, and L, on the basis of mathematical ability, where H represented the top third of this ability, M the middle third, and L the bottom third.

The total sample was also divided into m (usually 4) subsets on the basis of the classifying variable. In our example, the four subsets ranged from C_1, students whose teachers had had fewer than 4 years of teaching experience, to C_4, students whose teachers had had 13 or more years of teaching experience.

These two partitions of each of the four sets of students determined a 3-by-4 matrix for each of the four sets of students, each student in a set being assigned to one of the cells of the matrix according to the student's ability and his teacher's teaching experience.

Various descriptive statistics were then computed for each cell of the matrix. Whenever a statistically significant result was obtained, the results of the statistical analysis were reproduced in the appropriate position in NLSMA Reports Nos. 21-25. The analyses of the teacher variables appear in NLSMA Report No. 23, along with graphical representations of the results. The reader is referred to these volumes for the details of the statistical procedures and for the results. For present purposes it is enough to indicate that the statistics were sufficient to determine whether there was a main effect, i.e., whether achievement on the mathematics scale was significantly different for different values of the classifying variable. In addition, the analyses located interactions, i.e., cases where the effect of the classifying variable was significantly different at one ability level from its effect at another level.

For example, it turned out that for fourth grade girls using the SMSG textbook, scores at the end of the year on a test (X102) of division of whole numbers were affected by the amount of teaching experience their teachers had had. The more experience, the higher the scores. And this was true at all three ability levels, i.e., there was no inter-action effect.

On the other hand, at the end of the fifth grade, for boys using a conventional textbook, the number of years of teacher experience had no overall effect on scores on a geometry test (X305), but there was an interaction with ability level. High and middle ability boys profited from increased teacher experience, while low ability boys did worse as teacher experience increased.

We present in Table 3.1 and 3.2 summary statistics on main effects and interactions for each of the teacher background variables measured in NLSMA. We note first, however, that altogether there were 52 mathe-matics scales utilized in the C-Variable analysis. Since there were four sets of students analyzed for each of these, there were 4 × 52 = 208 possible main effects and as many possibilities for interactions. In some cases, however, not all the mathematics scales were used. Thus, for example, scales measured in grades 4, 5, and 6 were not used in the analysis of the teacher variable T17: Undergraduate Major in Mathematics. Too few of the elementary school teachers had majored in mathematics to allow a meaningful statistical analysis.

These summary statistics for teacher background variables are shown in Table 3.1. The names of the teacher variables and an explanation of the meaning of the positive and negative main effects are shown in Table 3.2.

Table 3.1

Summary Statistics for Teacher Background

Scale	Main Effects			Number of Main Effects				Interactions	Cognitive Level (in %)				School Level (in %)		
	No.(%)	+	−	SMSG	CONV	BOYS	GIRLS	No.(%)	Ct	Ch	Ap	An	EL	JH	SH
T02	53(25)	47	6	24	29	25	28	16(8)	29	25	38	13	15	26	33
T03	31(15)	23	8	5	26	18	13	25(12)	19	18	10	10	9	23	8
T04	53(25)	35	18	24	29	21	32	16(8)	45	23	25	17	26	31	13
T05	37(18)	21	16	13	24	18	19	16(8)	29	19	3	13	21	21	4
T06	61(29)	49	12	15	46	27	34	25(12)	46	25	34	15	23	33	31
T07	47(23)	36	11	22	25	26	21	19(9)	38	14	19	23	29	19	19
T08	49(24)	29	20	14	35	25	24	30(14)	38	19	9	25	24	29	13
T09	37(18)	27	10	17	20	15	22	16(8)	27	13	13	19	8	27	17
T10	43(21)	30	13	19	24	17	26	23(11)	29	16	25	15	11	23	31
T11	31(15)	17	14	10	21	11	20	11(5)	15	15	13	15	10	13	25
T12	21(10)	12	9	5	16	9	12	18(9)	19	8	9	6	8	7	19
T17	28(21)	19	9	10	18	12	16	11(8)	25	27	20	18	--	23	19
T18	39(19)	26	13	18	21	19	20	12(6)	31	15	13	15	19	15	23

Explanation: For teacher scale T02, there were 53 out of a possible 208 (25%) main effects. Of these, 47 were positive and 6 were in the opposite direction. See Table 3.2. Also, of these the SMSG students accounted for 24 and the conventional students for 29. Boys accounted for 25 and girls for 28. There were 16 out of a possible 208 (8%) interactions. Among the scales at the cognitive level of computation main effects occurred in 29% of the possible cases, for the comprehension level it was 25%, for application level it was 38%, and for the analysis level 13%. Scales administered to elementary school students had significant main effects 15% of the time, while for junior and senior high school students the percentages were 26% and 33%.

41

Table 3.2

Characterizations of Teacher Background Variables

TO2   Number of Years Teaching. A positive main effect indicates that more years of teaching are associated with higher student achievement.

TO3   Highest Academic Degree. A positive main effect indicates that students of teachers with an MA degree had higher scores than students whose teachers had only a BA degree.

TO4   Academic Credits Beyond BA. A positive main effect indicates that more graduate academic credits are associated with higher student achievement.

TO5   Math Credits Beginning with Calculus. A positive main effect indicates that more mathematics credits at the level of calculus or beyond are associated with higher student achievement.

TO6   Credits in Math Methods. A positive main effect indicates that more credits in mathematical methods courses are associated with higher student achievement.

TO7   In-service or Extension Courses. A positive main effect means that more in-service or extension courses are associated with higher student achievement.

TO8   Other Preparation in Last 5 Years. A positive main effect means that more teacher activities within the past 5 years are associated with higher student achievement.

TO9   Sex. A positive main effect indicates that students with female teachers scored higher than students with male teachers.

T10   Age. A positive main effect indicates that students of older teachers scored higher than students of younger teachers.

T11   Current Marital Status. A positive main effect indicates that students of single teachers scored higher than students of married teachers.

T12   Children. A positive main effect indicates that students whose teachers had children scored higher than students whose teachers were childless.

T17   Math as a Major or Minor. A positive main effect indicates that students whose teachers had an undergraduate major or minor in mathematics scored higher than students whose teachers did not. (Only junior and senior high school students were included in the analysis of this variable.)

T18   Major Field. A positive main effect indicates that students whose teachers were at the math end of the scale--math, science/tech, liberal arts, education, elementary education--scored higher than students whose teachers were farther to the right of the scale.

Comments

1. The teacher characteristics specified by the variables listed
in Table 3.2,on the one hand,are easy to obtain information about and,
on the other hand, are straightforward and require no exercise of judge-
ment, unlike "enthusiasm" and "warmth." Also, each of these characteristics has, at one time or another, been held essential, by someone,
for effective teaching.

2. The summary statistics displayed in Table 3.1 do not support
the claims that any of these characteristics are powerful indicators of
teacher effectiveness. Even the strongest of these variables, T06:
Credits in Mathematics Methods, has a significant positive relationship
to mathematics achievement only 24 percent of the time (49 out of 208),
and another 6 percent of the time (12 out of 208) its effect is negative.

3. For each of the 13 variables, the number of main effects is
smaller for the modern (SMSG) students than for the conventional students.

4. Main effects for boys are somewhat fewer than for girls, but
the difference is probably not significant.

5. Although the statistics are not included in Table 3.1, the
distributions of interactions between curriculum types and between sexes
are similar to the distribution of main effects.

6. How can we explain the fact that of the 16 negative main effects for T05, only one involved students in modern curricula?

7. There may be other interesting patterns in the full set of
statistics in NLSMA Report No. 23. The reader is urged to inspect this
volume.

The NLSMA Advisory Panel was not particularly surprised at the
results just reported since anecdotal evidence had been available to the
effect that teacher background variables were not strongly related to
student achievement. The panel had thought, however, that it was important to check this impression on the large sample of teachers which
NLSMA had measured.

B) Affective Variables [31]

A number of different aspects of teacher opinions about mathematics,
students, themselves, etc. have been correlated with student mathematics

learning. In most of these studies, the correlation turned out to be rather low. Teacher attitudes seem to have little effect on student learning.

Since these findings go against the opinions of many mathematics educators, it is appropriate to review here the NLSMA data. The NLSMA data bank in this area is probably larger than the combined data banks of all the other 31 studies I have located.

The NLSMA Advisory Panel also had much higher hopes for measures of teacher attitudes toward various aspects of the teaching profession than they did for background variables as predictors of student mathematics achievement. Teacher attitude measures were gathered by means of a questionnaire sent to each of the teachers involved in the first three years of the NLSMA project. Returns were received from slightly more than 60 percent of the teachers, a percentage that is reasonably high for a mail questionnaire.

From the questionnaire, seven different, and seemingly independent, attitude scales were extracted. The questionnaire, the separate scales, and descriptive statistics for each scale are all found in NLSMA Report No. 9, to which the reader is referred.

These descriptive statistics, incidently, refute a commonly held belief that elementary school teachers dislike mathematics and are afraid of it. Admittedly, the population of teachers involved in NLSMA was not representative of the total population, but at least for this large set of teachers the average attitude toward mathematics was at worst neutral.

Table 3.3 provides summary statistics similar to those in Table 3.1, on the effects of these teacher attitudes on student achievement. Table 3.4 describes these teacher attitude variables.

Table 3.3

Summary Statistics for Teacher Attitude Variables

Scale	Main Effects			Number of Main Effects				Interactions	Cognitive Level(in %)				School Level(in %)		
	No.(%)	+	-	SMSG	CONV	BOYS	GIRLS	No.(%)	Ct	Ch	Ap	An	EL	JH	SH
T20	47(23)	31	16	26	21	14	33	25(12)	33	21	19	13	25	24	15
T21	49(24)	36	13	22	27	28	21	28(13)	50	16	9	17	19	26	25
T22	39(19)	22	17	17	22	26	13	51(25)	31	11	19	17	20	18	17
T23	63(30)	31	32	23	40	31	32	35(17)	42	25	38	21	23	31	40
T24	43(21)	31	12	16	27	21	22	37(18)	23	24	9	19	16	25	19
T25	50(24)	33	17	25	25	26	24	30(14)	31	24	22	17	16	26	31
T26	57(27)	34	23	16	41	30	27	24(12)	52	20	13	23	26	29	25

Table 3.4

Characterizations of Teacher Attitude Variables

T20   Theoretical Orientation. A positive main effect indicates that teacher concern with insight and understanding, as opposed to rote memory, was associated with higher student achievement.

T21   Concern for Students. A positive main effect indicates that *less* empathetic concern for the social and emotional development of students was associated with higher student achievement.

T22   Involvement in Teaching. A positive main effect indicates that higher satisfaction with and interest in teaching, on the part of the teacher, was associated with higher student achievement.

T23   Nonauthoritarian Orientation. A positive main effect indicates that the teacher's belief in a nonauthoritarian approach, as opposed to strict discipline, was associated with higher student achievement.

T24   Like vs. Dislike. A positive main effect indicates that the greater the degree to which the teacher liked and valued mathematics, the higher the student's achievement.

T25   Creative vs. Rote. A positive main effect indicates that the higher the teacher's belief that learning mathematics is primarily a creative process, as opposed to a memory task, the more it was associated with higher student achievement.

T26   Need for Approval. A positive main effect indicates that the degree to which a teacher endorsed desirable statements about his competence as a teacher was associated with higher student achievement.

Comments

1. As in the case of the background variables, these attitude variables do not seem to have a strong influence on teacher effectiveness. Even the strongest of these variables, T21, had a positive main effect in only 17 percent of the possible cases.

2. For five of the seven variables, there were more main effects for students in a conventional curriculum than for the modern students, and for one of the two remaining variables there was a tie.

3. There seems to be no significant difference between boys girls as far as the distribution of main effects goes.

4. Again, the low cognitive level scales generally show more main effects than the scales at higher levels.

5. Again, the distribution of main effects across grade levels shows no major discrepancies.

6. Another analysis was carried out in which student achievement was compared for students whose teachers had been low for two consecutive years on a particular teacher variable, students whose teachers had been low one year but high the other, and students whose teachers had been high on the variable both years. The results of this analysis are also reproduced in NLSMA Report No. 23. No clear-cut patterns in the results emerged.

These analyses of NLSMA data merely investigated the relationships between teacher variables and student achievement. While they are enough to demonstrate that these teacher variables did not contribute very much to measures of teacher effectiveness, no direct measures of teacher effectiveness were calculated. Another analysis of the NLSMA data, in which such measures were calculated, may therefore be of interest.

Geeslin and I carried out a different kind of analysis of the NLSMA teacher variables, both background and attitude together, which we reported in NLSMA Report No. 28. We used stepwise regression in order to determine the relative importance of these variables for teacher effectiveness. The teacher sample consisted of all those teachers, for the first year of the study, who had returned the Teacher Opinion Questionnaire.

These teachers were first stratified by grade level, since it was possible that the characteristics of effective teachers could be different for different grade levels, and then by sex, since it was possible that the characteristics of effective female teachers might differ from those of effective male teachers.

A number of mathematical and psychological scales that had been administered at the beginning of the year were chosen as covariates for the measurement of teacher effectiveness and were also used to divide the students into a high ability (H) subset and a low ability (L) subset. The students were also stratified by sex. Finally, another stratification of the students was based on the textbook used during the academic year. The textbooks were classified either as conventional or as modern. Thus there were 3 (grade level) $\times$ 2 (teacher sex) $\times$ 2 (student ability level) $\times$ 2 (student sex) $\times$ 2 (textbook type) = 48 subsets of students. Each subset was analyzed separately.

For each grade level two end-of-year mathematics scales were chosen. One was at the cognitive level of computation (Ct) and the other at the cognitive level of comprehension (Ch). For each teacher two effectiveness scores were computed, for each student sex, on each of the two criterion measures. One score was the teacher's effectiveness in teaching high ability students computational skills, EFF-Ct-H. Of course, in computing this score all comparable (i.e., same sex and same textbook type) teachers were involved. Similarly, three other effectiveness scores were computed, EFF-Ct-L, EFF-Ch-H, and EFF-Ch-L.

Finally, an attempt was made to see if effectiveness for high ability students was similar to effectiveness for low ability students. For this purpose, the measures (EFF-Ct-H) - (EFF-Ct-L) and (EFF-Ch-H) - (EFF-Ch-L) were tested to see if they were significantly different from zero and, if so, how these measures were affected by teacher variables. Also, to see how effectiveness in teaching computation and in teaching comprehension compared, the measures (EFF-Ct-H) - (EFF-Ch-H) and (EFF-Ct-L) - (EFF-Ch-L) were investigated.

Once the stratification of teachers and students had been carried out and the various effectiveness measures listed above had been computed, the next step was to investigate the influence of teacher variables on these measures. The emphasis was on the teacher variables,

T20, . . . T26, so only four of the teacher background variables were included, T02, T12, T17, and T18. All these variables together were used in a stepwise regression analysis for each of the eight effectiveness measures for each of the 3 (grade level) × 2 (teacher sex) × 2 (text type) × 2 (student sex) = 24 sets of teachers. (Whenever it was statistically appropriate, the analyses for male and female students were combined.)

The stepwise regression analysis first picked out that teacher variable most highly correlated with the effectiveness measure being studied. Next, it picked out that one of the remaining teacher variables most highly correlated with the effectiveness measure after the first teacher variable had been partialled out. The program proceeded in this step-by-step fashion until the addition of another teacher variable did not significantly increase the amount of variance in the effectiveness measure accounted for by the teacher variables already chosen.

The full results of these analyses are reproduced in NLSMA Report No. 28. Here we present only some summary statistics and some comments based on the full report. Table 3.5 gives some impression of the frequency and magnitude of the teacher effects on the effectiveness scores.

Table 3.5

Summary Statistics for Teacher Effectiveness Study

	Grade 4	Grade 7 Modern	Grade 7 Conventional	Grade 10
Significant amount of variance accounted for: frequency	25%	63%	31%	13%
Significant amount of variance accounted for: average for significant cases only	12%	5%	37%	24%
Amount of variance accounted for: range for significant cases only	4% - 44%	2% - 9%	14% - 46%	7% - 36%

Note: The results, at grade levels 4 and 10, for the modern and the conventional textbooks were quite similar and have been amalgamated here.

The following comments are based not just on this table but also on the full report.

Comments

1. Significant relationships between teacher variables and effectiveness scores were not frequent, appearing in fewer than 30 percent of the possible cases.

2. Even in cases where there was a significant relationship between the teacher variables and the effectiveness measures, the amount of variance in the latter accounted for by the former was rather small. It must be remembered that the figures in the second row of Table 3.5 are the averages for the significant cases only.

3. Even after taking into account the fact that there were seven affective teacher variables and only four teacher background variables, the former emerged as the leading terms in the regression equation (i.e., the ones most highly correlated with the effectiveness scores) more than three times as often as the latter.

4. The most frequent affective regression terms were different for the three grade levels. For the fourth grade teachers, T25 was the teacher variable which appeared most frequently as the one most highly correlated with teacher effectiveness. For the seventh grade teachers, T24 appeared most frequently, and for the tenth grade teachers, T26 appeared most frequently.

5. There was no obvious pattern in the distribution of significant relationships between effectiveness with high ability students and with low ability students.

6. For the fourth and tenth grade teachers, all but one significant relationship involved a comprehension measure. For the seventh grade teachers, no such pattern emerged.

There were some byproducts from this investigation. For one, the printout for the stepwise regression analysis also provided the standard deviation of all the effectiveness scores. None of these were close to zero. There was substantial variance in all of the scores. Thus we have hard evidence from this investigation that teachers do differ in effectiveness.

50

C) Knowledge of Mathematics [17]

It is widely believed that the more a teacher knows about his subject matter, the more effective he will be as a teacher. The empirical
literature suggests that this belief needs drastic modification and in
fact suggests that once a teacher reaches a certain level of understanding of the subject matter, then further understanding contributes nothing
to student achievement. Such at least is the most reasonable interpre-
tation of a large study which I carried out (12) with teachers of ninth
grade algebra. Most of the other studies in this set can be interpreted
in the same way.

It is encouraging that there have been so many studies in this area.
Most teachers are reluctant, understandably, to submit to tests of their
grasp of subject matter. That as many as 17 sets of teachers were
willing to do so is a tribute to the profession.

D) Test Information [15]

This is a miscellaneous set of studies whose sole common characteristic is that information on a teacher characteristic which might
affect student learning was obtained by means of a test of something
other than teacher knowledge or teacher attitudes. No striking findings
emerged from any of these studies, but a careful review of them is probably indicated.

E) Stability of Effectiveness [6]

In addition to the studies reviewed by Rosenshine (11), I have located a few more which deal specifically with mathematics teaching.
They do little to contradict Rosenshine's conclusion that teacher effectiveness does vary over time.

In fact, strong support for this conclusion was obtained in the
analysis of NLSMA data which Geeslin and I carried out. We discovered
that a substantial number of teachers who had been teaching fourth grade
NLSMA students during the first year of the study were teaching fifth
grade NLSMA students during the second year. Similarly, a number of
the seventh grade teachers were again involved in the study with eighth
graders during the second year. We accordingly computed (using all the
fifth grade and eighth grade second year students) the same kind of

effectiveness scores for these teachers. Table 3.6 shows the correlations between the year one and the year two effectiveness scores.

Table 3.6

Correlations Between Effectiveness Scores
for Two Consecutive Years

	Grade 4 and Grade 5	Grade 7 and Grade 8
EFF-Ct-H	.35	.15
EFF-Ct-L	.25	.23
EFF-Ch-H	.01	.28
EFF-Ch-L	.06	.14

The maximum percentage of variance in the second year effectiveness scores accounted for by the first year scores was 12 percent. Thus, these effectiveness scores were not very stable.

Student Ratings [15]

As was indicated above, ratings of teachers, whether by administrative personnel, peers, or students, bear little relationship to teacher effectiveness as measured by student achievement. Nevertheless, recent concern, particularly on the part of college students, with the quality of the instruction they were receiving led to a vast increase in the amount of formal evaluation of faculty by students, expressed in the form of student ratings of their teachers. This in turn led to the appearance of a substantial number of studies of various aspects of this rating process. For those interested in pursuing these studies, an annotated bibliography prepared by Coley (13) will be helpful.

Only a small number of the studies were concerned with mathematics teachers. The results of these studies were mixed, but, taken as a whole, they do not conflict with the first sentence of the preceding paragraph.

There are many questions in connection with the grading practices of teachers, including mathematics teachers. A general discussion has been prepared by Warren (14).

The few studies gathered together here do little to answer these questions, but they do suggest strongly that teacher assigned grades do not correlate very well with student achievement as measured by objective tests and that, in addition, girls usually receive higher grades than boys.

However, until we have a clearer understanding of how we should interpret and use grades, the information available in these studies is of very little use to us.

Teacher Training Programs

The changes in the mathematics programs which were investigated in the sixties focused renewed attention on both pre-service and in-service programs for teachers and resulted in a number of evaluations of such programs, but almost exclusively at the elementary school level.

A) Elementary In-Service Programs [7]

These seven studies evaluated new in-service programs in terms of their effects on the teacher participants. Other programs which were evaluated in terms of the achievement of the teacher participants' students are taken up in Chapter Seven.

These studies were probably too few in number to allow us to derive any useful guidelines for future programs for in-service teachers, particularly since the criterion measures used in their evaluations are of doubtful validity, as the preceding discussions in this chapter have indicated. Nevertheless, anyone planning an in-service program for mathematics teachers in the elementary school should review these studies with care.

B) Elementary Pre-Service Programs [28]

A variety of aspects of pre-service courses have been studied, including the relationship between methods and content courses and use of elementary student materials in such courses. A detailed review of these studies might be of value, but a cursory inspection of the studies does not suggest any striking findings.

Summary Comments

Because of its widespread importance, the concept of teacher effectiveness has received a great deal of study. As a result, there exists a large reservoir of factual information about teachers and, in particular, about their effects on student learning.

We are more fortunate in this particular area of mathematics education than we are in some other areas in that, thanks to some substantial reviews of the empirical studies, much of the information about teachers has been pulled together, organized, and made easily available.

Probably the most important generalization which can be drawn from this body of information is that many of our common beliefs about teachers are false, or at the very best rest on shaky foundations. Thus, for example, there are no experts who can distinguish the effective from the ineffective teacher merely on the basis of easily observable teacher characteristics. Similarly, the effects of a teacher's subject matter knowledge and attitudes on student learning seem to be far less powerful than most of us had realized.

Consequently, we have to admit that we do not at present know of any way of selecting, in advance, the effective teachers or of knowing whether a particular teacher preparation program does indeed produce effective teachers.

In any case, it is clearly incumbent on anyone who wishes to participate in discussion of teacher effectiveness or teacher education to become familiar with the empirical literature which I have outlined in this chapter.

My overall reaction to the mass of information about teachers which is available to us is one of discouragement. These numerous studies have provided us no promising leads. We are no nearer any answers to questions about teacher effectiveness than our predecessors were some generations ago. What is worse, no promising lines of further research have been opened up. Evidently our attempts to improve mathematics education would not profit from further studies of teachers and their characteristics. Our efforts should be pointed in other directions.

Bibliography

1. Ryans, D. G. *Characteristics of Teachers: Their Descriptions, Comparisons, and Appraisal*. Washington, DC: American Council on Education, 1960.

2. Barr, A. S., et al. Wisconsin Studies of the Measurement and Prediction of Teacher Effectiveness: A Summary of Investigations. *Journal of Experimental Education*, Vol. 30 (1961) pp. 1-155.

3. Morsh, J. E., and Wilder, E. W. Identifying the Effective Instructor: A Review of the Quantitative Studies, 1900-1952. ED 044 371.

4. Rosenshine, B. Teaching Behaviors and Student Achievement. International Association for the Evaluation of Educational Achievement, IEA Studies No. 1. London: National Foundation for Educational Research in England and Wales, 1971.

5. Kulik, J. A., and McKeachie, W. J. The Evaluation of Teachers in Higher Education. In: Kerlinger, F. N., (Ed.), *Review of Research in Education*, Vol. 3. Otasca, IL: F. E. Peacock Publishers, 1975.

6. Popham, W. J. Performance Tests of Teaching Proficiency: Rationale, Development and Validation. *American Educational Research Journal*, Vol. 8 (1971) pp. 105-117.

7. Baral, L. R. The Evaluation of Teachers. ED 089 478.

8. Blount, G. Teacher Evaluation: An Annotated Bibliography. ED 093 033.

9. Canadian Teachers' Federation. Teacher Evaluation. ED 110 447.

10. Scott, C. S., and Thorne, G. Assessing Faculty Performance: A Partially Annotated Bibliography. ED 093 187.

11. Rosenshine, B. The Stability of Teacher Effects Upon Student Achievement. *Review of Educational Research*, Vol. 40 (1970) pp. 647-662.

12. Begle, E. G. Teacher Knowledge and Student Achievement in Algebra. SMSG Report No. 9. ED 064 175.

13. Coley, R. J. Student Evaluation of Teacher Effectiveness. ED 117 194.

14. Warren, J. R. The Continuing Controversy over Grades. ED 117 193.

Kane, R. B.   Attitudes of Prospective Elementary School Teachers Toward
Mathematics and Three Other Subject Areas. *Arithmetic Teacher,*
Vol. 15 (1968) pp. 169–175.

Gibney, T. C., Ginther, J. L., and Pigge, F. L.   A Comparison of the
Number of Mathematics Courses Taken by Elementary Teachers and Their
Mathematical Understandings. *School Science and Mathematics,*
Vol. 70 (1970) pp. 377–381.

Shim, C.   A Study of the Cumulative Effect of Four Teacher Characteristics on the Achievement of Elementary School Pupils. *Journal of
Educational Research,* Vol. 59 (1965) pp. 33–34.

Phillips, R. B., Jr.   Teacher Attitude as Related to Student Attitude
and Achievement in Elementary School Mathematics. *School Science
and Mathematics,* Vol. 73 (1973) pp. 501–507.

Eisenberg, T. A.   Begle Revisited: Teacher Knowledge and Student
Achievement in Algebra. *Journal for Research in Mathematics Education,* Vol. 8 (1977) pp. 216–222.

Gordon, M.   Mathematics Presentation as a Function of Cognitive/Per-
sonality Variables. *Journal for Research in Mathematics Education,*
Vol. 8 (1977) pp. 205–210.

Acland, H.   Stability of Teacher Effectiveness: A Replication.
Journal of Educational Research, Vol. 69 (1976) pp. 289–292.

Bryson, R.   Teacher Evaluations and Student Learning: A Reexamination.
Journal of Educational Research, Vol. 68 (1974) pp. 12–14.

Knafle, J. D.   The Relationship of Behavior Ratings to Grades Earned by
Female High School Students. *Journal of Educational Research,*
Vol. 66 (1972) pp. 106–110.

Hunkler, R.   An Evaluation of a Short-Term In-Service Mathematics Program for Elementary School Teachers. *School Science and Mathematics,* Vol. 71 (1971) pp. 650–654.

Litwiller, B. H.   Enrichment: A Method of Changing the Attitudes of
Prospective Elementary Teachers Toward Mathematics. *School Science
and Mathematics,* Vol. 70 (1970) pp. 345–350.

4

Curriculum Variables

I use the word "curriculum" to refer to the mathematical objects which we intend our students to absorb and incorporate into their cognitive structures and to the content of the instructional program devoted to these mathematical objects. Some educators use the word more broadly. For them "curriculum" refers not merely to the content of the instructional program but also to the manner in which the instruction is administered. This aspect of mathematics education I postpone to Chapter Seven.

In this chapter I consider those educational variables that are directly concerned with the curriculum. These variables are very important for the improvement of mathematics education, for, as will become evident, they are easy to manipulate and to incorporate in our instructional programs whenever they turn out to be helpful.

Adjunct Questions [5]

In 1966, Rothkopf (1) reported that student learning from prose could be increased by inserting questions at regular intervals throughout the prose passage. In addition, he reported that learning was increased more by questions placed after, rather than before, the exposition which provided their answers.

A considerable amount of further research has been carried out on the effects of these "adjunct" questions, and a review of the literature up to 1972 was prepared by Ladas (2). Most of the research used

prose material not relevant to mathematics education, but five studies
looked at the effects of adjunct questions in mathematics text materials.
Although one of them did not develop any significant advantage for adjunct questions, the others did. Further research on this topic is
needed.

Advance Organizers [24]

Ausubel (3) has developed a theory of meaningful learning, one
aspect of which is the "advance organizer." In his theory, meaningful
learning takes place when the learner relates new material in a substan-
tive and non-arbitrary fashion to his already existing cognitive structure. Ausubel also postulates that this process is facilitated if
appropriate introductory materials, called organizers, are provided.
These organizers, in Ausubel's view, should be broad and at a higher
level of abstraction and generality than the new material to be learned
so that the organizers can subsume the new materials and bridge the gap
between the new materials and the existing cognitive structure.

A considerable number of empirical studies have been carried out
which together do much to substantiate Ausubel's postulate about advance
organizers. Most of these studies used learning materials which were
outside the domain of mathematics, but there have been some using mathematical materials. In these latter studies, however, the advance organizers do not always accord with Ausubel's characterization and in
some cases were less abstract than the new material to be learned.

Although about half the studies found that the advance organizers
had no significant effects, it should be noted that none of the significant effects were negative. The advance organizers never did any harm.
Hence it would make sense to investigate advance organizers more widely.

Alternative Treatments

We all know that many mathematical objects can be approached in
more than one way. Among mathematical operations, we recall that there
are a number of different algorithms for subtraction, for example. We

know that there are different approaches to the concept of whole number multiplication, one in terms of repeated addition and one in terms of rectangular arrays. And of course we are all aware that a theorem may well have more than one proof.

It should not be surprising that there have been numerous attempts to compare two or more approaches to a mathematical object to see if one approach was pedagogically better than another. The many mathematical objects which have been studied in this way could be categorized in many different ways. The particular categorization which I follow below may not be the best one, but it should be reasonably informative for most readers.

1. *Addition and Subtraction of Whole Numbers* [26]

These studies were split about evenly between studies of the basic concepts and studies of the algorithms.

The significant findings are not sufficiently numerous to provide us with much guidance as yet, but they clearly indicate that further coordinated research along these general lines could have a substantial payoff. Such research, however, should be based on a careful analysis of the work which has already been done.

2. *Multiplication and Division of Whole Numbers* [20]

The studies were split about evenly between comparisons of algorithms, or approaches thereto, and approaches to the basic concepts.

The comments above on addition and subtraction of whole numbers are equally applicable here.

3. *Fractions* [22]

About half the studies were devoted to division of fractions. The comments above on addition and subtraction of whole numbers apply equally here.

4. *Decimals* [2]

It is astonishing that there has been so little work in this area.

5. *Percent* [3]

   See comments on decimals.

6. *Algebra* [14]

   Since these studies, between them, ranged from junior high school to college, the coverage at any one level is too thin to permit conclusions. Further studies are indicated.

7. *Geometry and Measurement* [23]

   As in the case of algebra, these studies ranged over many grade levels, from elementary school to college. There are enough significant findings to suggest that further studies would be quite appropriate.

8. *Analysis* [11]

   Seven of these studies compared different ways of teaching the limit concept. A careful review of these studies will probably suggest that further comparisons are not needed.

9. *Aptitude-Treatment Interactions* [23]

   Students differ very substantially among themselves on a large number of cognitive psychological tests. It has been suggested that instruction ought to take advantage of these differences. Thus for example, it is plausible to expect that students high on verbal ability but low on spatial visualization would do better on a highly verbal treatment of a mathematical topic than on a highly graphical treatment, while for students high on spatial visualization but low on verbal ability, the results on the two instructional treatments would be reversed. If this were to happen, we would have an example of an "aptitude-treatment interaction," or ATI.

   Plausible as the ATI concept may seem, it does not often show up in a significant way, and some of the studies contradict the results of others. It is too early, however, to give up this topic.

10. *Miscellaneous* [9]

   A few studies fell under none of the above headings.

Comment

I conclude from a review of these studies that whenever there are two different ways of developing a mathematical concept or two different algorithms for the same mathematical operation, it is more likely than not that one will be more effective than the other, at least under some circumstances. However, it is rarely possible to decide on a priori grounds which will be the more effective, and under which circumstances, and we have to conduct empirical studies to make the determination. The work which has already been done furnishes us with a good start, but there is much that remains to be done along these lines.

NOTE. Some of the studies listed under the heading of New Topics below might well have been placed here.

Behavioral Objectives [27]

The question investigated under this heading is: "If students are given the behavioral objectives of an instructional unit before they start to study it, will their achievement after studying the unit be higher than if they had not been given the objective?"

A moderate amount of empirical research has been devoted to this question, and there exist two reviews of this research. The first, by Duchastel and Merrill (4) is not restricted to mathematics instruction. They report a total of 25 studies. In twelve of these, there were no significant findings. In another twelve, the provision of behavioral objectives to the learners significantly increased learning. In only one case did behavioral objectives prove to be deleterious to learning.

Only four of these studies involved the learning of any mathematical topics, and three of these had nonsignificant findings.

The other review was prepared by Walbesser and Eisenberg (5). There is a good deal of overlap between these two reviews, but Walbesser and Eisenberg gave references to nine additional studies, six of which, including one on a mathematical topic, found the provision of behavioral objectives to either improve learning or speed it up.

In addition to the studies reported in these two reviews, I have found additional references that bring the total number of empirical studies of the effects of behavioral objectives on student learning of mathematics to over two dozen. Some of these additional studies, incidentally, provided the behavioral objectives to the teachers rather than the students.

Almost half the studies reported significant effects. Only in one case did the provision of behavioral objectives have a negative effect on students. Further investigation of this variable is clearly called for.

Commercial Textbooks [5]

The work of the CEEB Commission on Mathematics drew a great deal of attention to the desirability of modifying the high school mathematics curriculum. The work of SMSG demonstrated that there was a substantial market for textbooks which moved in the direction of the Commission's recommendations. As a result, a substantial number of textbooks were developed through the normal commercial channels which followed the Commission's directions to some extent.

However, I was able to locate only five studies, outside of NLSMA, in which such textbooks were compared or evaluated. As in the case of textbooks developed by the various curriculum projects, discussed below, the findings of these studies were not very helpful. The reader is referred to the comments on the project textbooks, since they apply also to commercial textbooks.

It should be noted that some of the textbooks produced by curriculum projects during the sixties eventually were taken over by commercial publishers. None of these were involved in the above studies.

Drill [26]

During the twenties and thirties there were many studies and experiments carried out on the teaching and learning of rote computational skills, done almost entirely by means of drill. These activities continued, although at a reduced rate, when it became clear, as is discussed below, that drill should always be preceded by meaningful teaching of the algorithms to be drilled.

An excellent booklet for the classroom teacher about computational skills has been prepared by Suydam and Dessart (6).

Even though six of the studies which I have located indicated no significant value for drill, there have been enough studies showing a significant positive effect to strongly suggest further research aimed at pinning down more closely the optimal kinds, amounts, and timing of drill work.

It is interesting to note that another six of these studies were concerned with mental computation.

Formal Presentations [9]

The word "formal" is often used to describe mathematical programs, but it is ambiguous. To some writers it means "rigorous," to others it means "axiomatic," and sometimes it is used to describe a program in which a particular symbolism or terminology is rigidly insisted on.

Despite the strong feelings which the word "formal" often arouses, only a few studies have been done in which formality, in any of the above senses, was the major variable of interest. The results were mixed and certainly do not indicate that the presence (or lack of it) of formality is a strong curriculum variable.

Undoubtedly, many of the studies under other headings bear on this question, but until the term is made precise, not much more can be said.

Readers still interested in this question may wish to review NLSMA Reports Nos. 10 through 18. These provide a great deal of information on the mathematics achievement of students who used particular textbooks over periods of one or more school years. If the reader wishes to classify some of these textbooks as being more formal than others, then the NLSMA data may throw some light on the effects of "formality" on student achievement.

According to the analysis, in Chapter One, of mathematical objects, these form a hierarchy in the sense that, except for the beginning ones, the specification of any mathematical object involves other, prerequisite mathematical objects. The latter come below the specified object in the hierarchy.

It seems plausible then that to learn a mathematical object a student would first have to learn each of the objects strictly below it in the hierarchy. But, of course, the question immediately arises as to whether this plausible argument is actually valid. This leads to the notion of a *learning hierarchy* of mathematical objects. Two definitions have appeared in the literature. Mathematical object A is below (in the learning sense) mathematical object B if: (a) learning A makes it easier to learn B, or (b) B cannot be learned without first learning A. Most of the research uses the second definition.

The first work on this question seems to have been done by a psychologist, R. Gagné, who came at the matter from a slightly different direction. He was interested in mathematical tasks and what was required to learn how to do them. He asked the question: "Given a task and directions what would the student need to know how to do in order to learn how to do the task?" Gagné would then ask the same question about each of the tasks, prerequisite to the final task, which were identified by the first question. This would lead to tasks at a still lower level of the learning hierarchy, and the process could be continued until only simple basic tasks, ostensibly well within the capacity of any learner, were reached.

Now, given a hierarchy constructed by experts by some rational procedure, as above, it is still necessary to ask whether it is indeed a learning hierarchy or, in another terminology, whether the rational hierarchy is *valid*. The usual procedure is to construct an instructional unit which teaches all the objects (or tasks) in the rational hierarchy in a sequence such that no object is taught until all the objects strictly below it have been taught. At the conclusion of the unit, tests for each of the objects in the hierarchy are administered, and the results are inspected to see if condition (a) or condition (b) above is satisfied for each appropriate pair of objects.

It turns out that it is not always the case that a rationally constructed hierarchy is indeed a valid one. However, one or two revisions usually have been enough to produce a hierarchy near enough to validity for all practical purposes.

Most of the studies reported so far have been this kind of validity study. In some cases, however, a rational hierarchy was used to construct an instructional unit, and the effectiveness of the unit as a whole was all that was evaluated.

Further work on learning hierarchies would seem to be indicated. In addition to pure validation studies, some of the results to date suggest that it would be worth looking at comparisons of validation studies across grade levels or across ability levels.

It should also be mentioned that this topic overlaps strongly with the later topics of Meaningful Instruction and Sequence.

Three reviews of the general topic have appeared, authored by Briggs (7), Resnick (8), and Walbesser and Eisenberg (5).

Meaningful Instruction [36]

It is usually quite easy to distinguish between a rote and a meaningful presentation of a mathematical object. Nevertheless, there does not seem to appear in the literature a clear, widely accepted definition of "meaningful instruction." This quotation is typical: "Meaning is to be sought in the structure, the organization, the inner relationships of the subject itself." (20, p. 481)

The importance of meaning in mathematics education has been recognized for a long time, at least by some educators. But widespread acceptance of this importance can be traced back to two sources. The first source was the work of William A. Brownell, best typified in his now well-known study of subtraction in the third grade (9). This work and its implications were recognized primarily by those concerned with elementary school mathematics and did not come to the attention of university mathematicians, in particular, until long after the second source had been widely publicized. This second source was the set of recommendations

67

put forth by the Commission on Mathematics of the College Entrance Examination Board (10). Some educators had been urging increased attention to mathematical structure in high school mathematics education, but it was the Commission's Report which provoked widespread discussion and which provided guidance not only for the School Mathematics Study Group's curriculum development efforts but also those of many other smaller groups and individual textbook writers. It is interesting to note that most of the university mathematicians who worked on the high school and junior high school, and later on the elementary school curricula, were as much astonished as they were pleased to learn how thoroughly they had been anticipated by Brownell and his colleagues.

An excellent discussion and review of the pre-1960 work is to be found in a publication by Weaver and Suydam (11).

The three dozen references mentioned above include only those for which the word "meaningful" appears in the title of the article or else is clearly indicated, in the body of the report, as being the primary variable being studied. Of course, many other studies, such as those evaluating the SMSG textbooks, for example, can throw light on this variable.

This seems, on the basis of the references which I have inspected, to be an important variable. More than four-fifths of the references report that meaningful instruction was superior, in some way, to rote instruction. Further research should probably concentrate, not on this simple comparison, but rather on variations in ways of providing meaning in mathematics instruction, interactions with student ability levels and maturity levels, etc.

Modern Mathematics [28]

This is another ill-defined category. Many of the entries placed here could just as well have been placed under other headings, such as Meaningful Learning, for example. The rule of thumb which was used was to place reports here if the word "modern" appeared in the title or was stressed in the description of the study.

The general pattern of such reported results as were statistically significant sometimes showed somewhat of a decline in computational skill for students in modern programs but, at the same time, an increase in problem-solving achievement and in scores on other tests at higher cognitive levels. However, it is hard to derive any useful generalizations because such a wide variety of "modern" instructional programs was involved.

A curious sidelight is a report that an arithmetic subtest of a popular IQ test for children is no longer valid for those who have studied a modern elementary school mathematics curriculum.

New Topics

During the sixties a number of mathematical topics were introduced into instructional programs which were new, at least for the grade levels at which the programs were aimed. Usually, quite plausible arguments were advanced for introducing these topics, but fortunately numerous scholars felt that, despite the plausibility, it was important to assess the validity of these introductions. A substantial number of such validity studies have been carried out, enough so that it seems worthwhile to group them by topic.

It should also be mentioned that there have been numerous studies of knowledge about some of these topics, in particular logic and probability, on the part of students not exposed to formal teaching of the topics. This is discussed in Chapter Five.

1. *Logic* [39]

About half of these studies investigated the effects of the study of a unit on logic on student achievement in mathematics. Only six studies were concerned with the usual high school geometry course.

The remainder of the studies looked at the feasibility of teaching logic to elementary or junior high school students or at the effects of a formal study of logic on critical thinking ability.

2. *Geometry* [29]

In addition to a number of studies on the effectiveness of transformational geometry at various grade levels, there have been studies of teaching coordinate, projective, and non-Euclidean geometry and also topology to elementary and secondary school students. Generally, it was found feasible to teach at least some aspects of the subject.

3. *Mathematical Objects* [71]

The Report of the Commission on Mathematics of the College Entrance Examination Board recommended that attention be paid to concepts as well as to skills. Accordingly, most of the curriculum projects of the sixties did attempt to teach students the basic mathematical concepts.

Two teaching procedures were used. The "discovery" method aimed at having students develop the concepts themselves, perhaps with guidance from the teacher. Research involving discovery teaching is summarized in a later chapter.

The expository method of teaching concepts provided students with an explicit definition of the concept. What is needed in addition to the definition—such as illustrative examples and non-examples, practice problems, emphasis on prerequisite concepts—has been the subject of some research. No clear generalizations have yet emerged from these studies.

Research studies on the learning of algorithms for mathematical operations are included under the heading "Alternative Treatments."

4. *Non-Decimal Number Bases* [14]

Most of the curriculum developers of the sixties felt that it was important to lead students to see why arithmetic algorithms worked. To do this, it was essential to have students thoroughly understand the Hindu-Arabic decimal place-value system of numeration. It was generally believed that a brief study of non-decimal bases would help students to gain this understanding.

A moderate amount of research has been devoted to the validity of this last belief. On the whole, this research suggests that non-decimal number bases do not contribute significantly to understanding the decimal place-value numeration system.

5. *Probability and Statistics* [11]

These topics have been advocated for inclusion in the mathematics curriculum for all students on the grounds that effective functioning in our society requires an appreciation of the power and of the limitations of statistical information. Some research has been done on the feasibility of including the rudiments of probability theory in the curriculum. Most of the results have been positive.

It should be mentioned that non-quantitative investigations of the feasibility of teaching probability and statistics to pre-college students have also been carried out on a fairly wide scale. The CEEB Commission prepared a senior level text (12) which was widely used and imitated. SMSG also prepared a number of elementary and junior high school probability units which saw widespread use, and the inclusion of probability in commercial textbooks at these levels is now common.

6. *Miscellaneous* [46]

A variety of empirical studies of the feasibility of various other topics--in algebra, analysis, measurement, computer-oriented mathematics, etc.--have been carried out. Inspection of these studies does not seem to suggest any useful or important generalizations.

One review article is available (13) covering research on the teaching of the metric system. It is curious that in this whole area of new topics there has appeared only this one review.

Preschool Programs [15]

All children in this country are required to attend school, beginning with grade one. Preschool programs, kindergarten or nursery school, are made available to some children but not to all. The effects of such programs on achievement in the regular elementary school are therefore of interest.

Although the studies I have managed to locate are not unanimous in their findings, they do suggest that preschool experience leads to better mathematics achievement in the early grades. However, it is not clear how long this advantage persists.

This heading refers not to the development of problem-solving skills, which will be treated in Chapter Nine, but rather to the use of problems as pedagogical tools in teaching mathematics.

It is sometimes asserted that the best way to teach mathematical ideas is to start with interesting problems whose solution requires the use of ideas. The usual instructional procedure, of course, moves in the opposite direction. The mathematics is developed first and then is applied to problems.

Unfortunately, we have very little factual information which bears on this question. Only two studies were found. The other three studies were concerned with such questions as the effects on mathematics achievement of the ordering and setting of problems to accompany text materials.

This is clearly an area in need of much more research. Problems play an essential role in helping students to learn concepts. Details of this role, and of the role of problems in learning other kinds of mathematical objects, are much needed.

Project Textbooks [118]

In addition to the School Mathematics Study Group, a number of other projects produced, during the sixties and early seventies, textbooks at both the elementary and the secondary level. Some, but not all, of these textbooks were eventually published commercially.

Lockard (14) has provided brief summaries of most of the science and mathematics projects.

Most of these projects carried out some evaluations of their products. In addition, in some cases, interested school systems did some empirical studies. And, of course, a number of doctoral students did comparisons of project textbooks with conventional texts for their dissertations. Also, other doctoral students investigated related questions, such as, for example, the effectiveness of a high school textbook as preparation for a college level calculus course.

In addition to these individually conducted empirical studies, a rather substantial comparison of three high school level projects was carried out by the Minnesota National Laboratory (15), under the direction of Professor Paul Rosenbloom. (This Laboratory also carried out a substantial amount of formative evaluation for SMSG.)

But the major investigation of the effectiveness of project textbooks was NLSMA, the results of which appear in Reports Nos. 10 through 18. Not all the projects were included in NLSMA since it started in 1962, before some of the projects had gotten under way. On the other hand, many of the textbooks included in the NLSMA analyses had been developed by individuals or small teams for commercial publishers in the customary fashion.

None of the statistically significant findings from this large array of studies was particularly startling. None of the textbooks turned out to be either a universal panacea or a complete disaster. Perhaps the most interesting findings were those of NLSMA which demonstrated clearly, at the upper elementary level, distinct patterns of achievement resulting from different textbook sequences, patterns which illustrated the independence of the different cognitive levels discussed in Chapter Two. Also, as the NLSMA results made clear, content differences in textbooks resulted in differences in student achievement.

But aside from mathematical structure and mathematical content, variables on which both project and commercial textbooks exhibited a fairly wide variation, no other curriculum variables were identified by any of these studies as being strong or important. One reason for this, perhaps, is that most of the studies, except for those done by NLSMA, used only standardized tests to measure achievement.

However, even if differences between textbooks had shown up which were significant practically as well as statistically, not much immediate use could have been made of these differences. Any two textbooks differ on so many variables that it would be almost impossible to trace the specific variables which cause a specific difference, and without knowing which variables make a difference, we do not know where to start to improve textbooks.

But all this work was far from wasted. Any sample of these studies which is well spread out with respect to time makes it clear that our research and evaluation techniques have dramatically improved over the past 15 or so years. We can now feel confident about being able to carry out a rigorous summative evaluation of any textbook or series. Those states which have a say in the choice of textbooks which their students may use no longer have any excuse for not carrying out empirical evaluations before making a choice.

Review [14]

Most teachers provide some review, after a certain amount of material has been covered, before going further. Obviously there are many possible variations in the amount, kind, and timing of review materials.

Such empirical studies in this area as have come to my attention clearly indicate that review can be beneficial and that a fairly substantial research program investigating the comparative effects of the possible variations could have useful payoffs.

Science [17]

It is well known that science has a powerful motivating effect on the learning of mathematics. At least, that is what many mathematicians and most scientists would have us believe. In addition, some scientists urge that mathematics and science, usually physics, be integrated in the curriculum, since the latter depends so strongly on the former.

We have very little empirical information which bears on either of these topics, and what little we do have is not very persuasive. In any case, the importance these days of mathematics in the biological and social sciences and in business and industry raises doubts about the wisdom of singling out physical science alone for such integration.

There is considerably more information available about the importance of mathematics to science education, although I have not systematically collected references to the empirical studies. In particular, success in high school mathematics seems to be a much better predictor of success in college science courses than success in high school science courses. My impression of the results of all these studies is that to improve science education, some of the time now devoted to science instruction should instead be used for more mathematics instruction.

Sequence [34]

Even when a number of mathematical objects are arranged in a hierarchy, it is usually possible to find many different sequences in which the objects can be taught without violating the hierarchical relations. However, while there is no logical reason for choosing one sequence over another, it may well be that one is pedagogically more effective.

A moderate number of empirical studies of several aspects of this topic have been carried out. The studies have been done at all levels, from kindergarten to college. In general, while statistically significant differences often show up in these studies, the actual differences are rarely large.

A useful review on this topic has been prepared by Heimer (16).

It should be noted that another kind of sequence study will be discussed in the next chapter in the section on programmed instruction.

Special Instructional Programs [91]

Two special problems have long been the concern of mathematics educators. What kind of mathematics program is best for the low achiever? What kind of mathematics program is best for the gifted student? The latter concern is discussed in a later chapter which takes up the topic of acceleration for gifted students.

The preparation of special instructional packages, including teacher training as well as curriculum units, for students doing poorly in mathematics received a great deal of support in the sixties. This happened primarily because of the national concern for improved education for minority and disadvantaged students. As a result of this concern, financial support was made available to local school systems to allow the development of special curricula aimed specifically at their own students. Numerous evaluation reports on these curriculum developments were prepared and became available through ERIC. At the same time, numerous doctoral students in mathematics education became interested either in the general problem or in nearby school system projects and based their dissertations on evaluations of the product of these projects.

No startling results issue from these reports. In a majority of cases, the special curriculum package did do some good, but the extent of the benefit was never very large. On the other hand, the special program very seldom did any harm.

Unfortunately, this rather large number of studies provides us with essentially no guidance for further work in this general area. There are two reasons for this. In the first place, most of the reports fail to describe in any detail the curriculum packages which were developed. Without a careful inspection of the actual curriculum materials, which would be difficult to obtain at this time and which, in the case of lectures to or discussions with classroom teachers, may not have been written down at all, it is impossible to sort out the various instructional variables involved and hence impossible to ascertain their relative importance.

Secondly, the student populations used in these studies were all quite localized, so it would be difficult to know which results could be generalized to other populations and how far they could be generalized.

Despite these negative comments, it would seem worthwhile to conduct as detailed a review of these studies as possible. Many concerned educators have devoted a lot of thought to the educational problems of minority and disadvantaged students. Their ideas should be assembled and sifted through. It is likely that some of them are worth wider evaluation.

Structure [21]

Does an emphasis on the structure of mathematical systems lead to improved student achievement in mathematics? In view of the great

emphasis placed on the commutative, associative, and distributive properties of arithmetic systems by the proponents of reform in mathematics
education, one would expect that there would be numerous factual studies
of the above question. However, there have only been a few in this
general area, and some of them were concerned with ways of discovering
how well the students had assimilated the structure rather than how well
they could use it.

Nevertheless, such evidence as has been gathered has all pointed to
structure as a facilitator of mathematics achievement. But it is clear
that further studies are called for.

Textbook Characteristics [11]

A small number of studies have been carried out on two physical
characteristics of textbooks: the use of color and the use of visually
different displays (e.g., horizontal vs. vertical presentation of addition problems, bar vs. circle graphs).

Color is used lavishly in elementary school textbooks and, con-
sidering the number of textbooks purchased each year in this country,
must account for a rather substantial part of the national educational
budget. It is therefore surprising that essentially no studies have
been made to see if the use of color, as opposed to black-and-white
illustrations and diagrams, does indeed result in improved learning
of mathematics. In particular, there is no evidence that it does so,
and the NLSMA results would suggest that the money now used to provide
color in textbooks might better be used for other purposes.

The relative effectiveness of different kinds of graphic displays
is clearly an area on which we ought to have more information.

Probably there are other textbook characteristics (e.g., type style,
size of page) which affect student learning, but I have not found any
empirical studies of them.

Verbal Variables [52]

A considerable amount of study has been devoted to the role of
verbal variables in mathematics education. But the research has been

spread out over a number of different kinds of questions, so that the research on any particular question is not yet deep enough to provide any definitive answer.

Among these topics studied were: the effects of verbalizing discovered mathematical objects; the effects on achievement of the reading level of textbooks; the effects on mathematics achievement of remedial reading programs, or of programs to develop mathematical vocabulary; the status of mathematical vocabulary knowledge among students; and measures of reading level.

There have been three useful reviews of the research in this area, one by Earps (17) and two by Aiken (18, 19).

This is clearly an area in which further study would have useful results.

Summary Comments

What lessons can we derive from all these studies of the effects of curriculum variables? There are a few.

1. A substantial number of curriculum variables have been identified and have been shown to be amenable to empirical study.

However, we have no way of knowing how complete our list is. It may be that we have already identified all those that are important, or, on the other hand, there may be many powerful variables not yet studied because no one has yet thought to do so.

2. The results on such of these curriculum variables as have been studied have been mixed. On the one hand, there are some--such as drill, meaningful instruction, and certain new topics such as probability-- where the results to date pretty clearly indicate that the variable is important. For these we need further studies to pin down the details and to estimate the relevant parameters.

For others, the results to date are not conclusive but have been positive often enough to suggest further study. Among these are adjunct questions, advance organizers, learning hierarchies, logic, review, and verbal variables.

Still others can probably now be dismissed as either useless or even harmful, such as formal treatments.

Together, these empirical results make it clear that empirical studies are important and can have a useful payoff. This is an especially important conclusion, since some of the results which have been found were unexpected and even fly in the face of conventional wisdom.

3. The studies of textbooks and of special programs, such as those for preschoolers and for disadvantaged or minority students, have been helpful in alleviating fears about new curricula and in holding down the over-enthusiastic proponents of these curricula. But for these results we did not need the very large number of studies which have been carried out.

On the other hand, as was pointed out above, the evaluation methodologies which have been developed during the last 15 to 20 years are powerful and useful and can be of great help to us in the future.

Nevertheless, these broad comparisons of different curriculum materials provide very little information that can be used to improve such materials or that can be used in theory building. The reason is that the materials being compared differ on so many different variables, including method variables as well as curriculum variables, that when different results are found we do not know which variables are responsible.

Except for formative and summative evaluations of new textbook materials and new special programs, empirical studies, if they are to be productive, should deal with only one variable at a time or be so designed that the effects of individual variables can be assessed.

4. The most important, and the most discouraging, observation coming from this survey is the lack of theoretical direction for these studies. True, the studies of advance organizers does derive from Ausubel's theory of meaningful reception learning, and the work on learning hierarchies does have a theoretical flavor, but in general these studies of curriculum variables are ad hoc and arbitrary. The particular questions addressed, many of them interesting and clearly relevant to the improvement of mathematics education, were not suggested by any broad theory about the learning of mathematics.

Until we have such a theory, or even better several of them, research on this part of mathematics education, and indeed on the other parts which are reviewed below, will have to continue in a hit-and-miss fashion and will be less efficient and less productive than we would wish.

Bibliography

1. Rothkopf, E. Z. Learning from Written Instructive Materials: An Exploration of the Control of Inspection Behavior by Test-like Events. *American Educational Research Journal*, Vol. 3 (1966) pp. 241-249.

2. Ladas, H. The Mathemagenic Effects of Factual Review Questions on the Learning of Incidental Information: A Critical Review. *Review of Educational Research*, Vol. 43 (1973) pp. 71-82.

3. Ausubel, D. P. *Educational Psychology: A Cognitive View*. New York: Holt, Rinehart and Winston, 1968.

4. Duchastel, P. C., and Merrill, P. F. The Effects of Behavioral Objectives on Learning: A Review of the Empirical Studies. *Review of Educational Research*, Vol. 43 (1973) pp. 53-69.

5. Walbesser, H. H., and Eisenberg, T. A. A Review of Research on Behavioral Objectives and Learning Hierarchies. Mathematics Education Reports. ERIC Information Analysis Center for Science, Mathematics, and Environmental Education. Ohio State University, (1972) ED 059 900.

6. Suydam, M. N., and Dessart, D. J. *Classroom Ideas from Research on Computational Skills*. Reston, VA: National Council of Teachers of Mathematics, 1976.

7. Briggs, L. J. Sequencing of Instruction in Relation to Hierarchies of Competence. (1967) ED 018 975.

8. Resnick, L. B. Hierarchies in Children's Learning. (1971) ED 064 668.

9. Brownell, W. A., and Moser, H. E. *Meaningful vs. Mechanical Learning: A Study in Grade III Subtraction*. Durham, NC: Duke University Press, 1949.

10. College Entrance Examination Board. *Program for College Preparatory Mathematics*. Report of the Commission on Mathematics. New York: CEEB, 1959.

11. Weaver, J. F., and Suydam, M. N. Meaningful Instruction in Mathematics Education. ED 068 329.

12. Commission on Mathematics. *Introductory Probability and Statistical Inference*. New York: College Entrance Examination Board, 1959.

13. Murphy, M. O., and Polzin, M. A. A Review of Research in the Teaching of the Metric System. *Journal of Educational Research*, Vol. 62 (1969) pp. 267-270.

14. Lockard, J. D. Science and Mathematics Curricular Developments Internationally, 1956-74: The Ninth Report of the International Clearinghouse on Science and Mathematics Curricular Developments. ED 106 112.

15. Ericksen, G. L., and Ryan, J. J. Secondary Mathematics Evaluation Project. A Study of the Effects of Experimental Programs on Pupil Achievement Observed During the First Three Years of the Project. ED 011 977.

16. Heimer, R. T. Conditions of Learning in Mathematics: Sequence Theory Development. *Review of Educational Research*, Vol. 39 (1969) pp. 493-508.

17. Earps, N. W. Reading in Mathematics. ED 036 397.

18. Aiken, L. R. Verbal Factors and Mathematics Learning: A Review of Research. *Journal for Research in Mathematics Education*, Vol 2 (1971) pp. 304-313.

19. Aiken, L. R. Language Factors in Learning Mathematics. *Review of Educational Research*, Vol. 42 (1972) pp. 359-385.

20. Brownell, W. A. When Is Arithmetic Meaningful? *Journal of Educational Research*, Vol. 38 (1945) pp. 481-498.

Mayer, R. E. Forward Transfer of Different Reading Strategies Evoked by Testlike Events in Mathematics Text. *Journal of Educational Psychology,* Vol 67 (1975) pp. 165-169.

Lesh, R. A., Jr., and Johnson, H. Models and Applications as Advanced Organizers. *Journal for Research in Mathematics Education*, Vol. 7 (1976) pp. 75-81.

Kratzer, R. O., and Willoughby, S. S. A Comparison of Initially Teaching Division Employing the Distributive and Greenwood Algorithms with the Aid of a Manipulative Material. *Journal for Research in Mathematics Education,* Vol. 4 (1973) pp. 197-204.

Bierden, J. E. Behavioral Objectives and Flexible Grouping in Seventh-Grade Mathematics. *Journal for Research in Mathematics Education,* Vol 1 (1970) pp. 207-217.

Miller, G. H. Theory or Practice in Arithmetic--Which Shall It Be? A Comparison Between a "Modern" Program and a "Modified Traditional" Program. *School Science and Mathematics,* Vol. 70 (1970) pp. 115-120.

Roberts, D. M., and Bloom, I. Mathematics in Kindergarten--Formal or Informal? *Elementary School Journal*, Vol. 67 (1967) pp. 338-341.

Phillips, E. R., and Kane, R. B. Validating Learning Hierarchies for Sequencing Mathematical Tasks in Elementary School Mathematics. *Journal for Research in Mathematics Education,* Vol. 4 (1973) pp. 141-151.

Shuster, A. H., and Pigge, F. L. Retention Efficiency of Meaningful Teaching. *Arithmetic Teacher,* Vol. 12 (1965) pp. 24-31.

Austin, G. R., and Prevost, F. Longitudinal Evaluation of Mathematical Computational Abilities of New Hampshire's Eighth and Tenth Graders, 1963-67. *Journal for Research in Mathematics Education,* Vol. 3 (1972) pp. 59-64.

McGinty, R. L. The Effects of Instruction in Sentential Logic on Selected Abilities of Second- and Third-Grade Children. *Journal for Research in Mathematics Education,* Vol. 8 (1977) pp. 88-96.

Stephens, L., and Dutton, W. H. The Development of Time Concepts by Kindergarten Children. *School Science and Mathematics,* Vol. 69 (1969) pp. 59-63.

Flanagan, S. S.   The Effects of SMSG Texts on Students' First Semester
   Grade in College Mathematics. *School Science and Mathematics,*
   Vol. 69 (1969) pp. 817-820.

Burns, P. C.   Intensive Review as a Procedure in Teaching Arithmetic.
   Elementary School Journal, Vol. 60 (1960) pp. 205-211.

Higgins, J. L.   Attitude Changes in a Mathematics Laboratory Utilizing
   a Mathematics--Through-Science Approach. *Journal for Research in
   Mathematics Education,* Vol. 1 (1970) pp. 43-56.

Willson, G. H.   Decimal-Common Fraction Sequence Versus Conventional
   Sequence. *School Science and Mathematics,* Vol. 72 (1972) pp. 589-
   592.

Lerch, H. H., and Kelly, F. J.   A Mathematics Program for Slow Learners
   at the Junior High Level. *Arithmetic Teacher,* Vol. 13 (1966)
   pp. 232--236.

Geeslin, W. E., and Shavelson, R. J.   Comparison of Content Structure
   in High School Students' Learning of Probability. *Journal for
   Research in Mathematics Education,* Vol. 6 (1975) pp. 109-120.

Weaver, J. F.   Pupil Performance on Examples Involving Selected Variations of the Distributive Idea. *Arithmetic Teacher,* Vol 20
   (1973) pp. 697-704.

Feliciano, G. D., Powers, R. D., and Kearl, B. E.   The Presentation of
   Statistical Information. *Audio-Visual Communication Review,*
   Vol. 11 (3) pp. 32-39.

Gilmary, S.   Transfer Effects of Reading Remediation to Arithmetic
   Computation when Intelligence is Controlled and All Other School
   Factors Are Eliminated. *Arithmetic Teacher,* Vol. 14 (1967)
   pp. 17-20.

Beauregard, R. J.   Construction and Validation of a Scale to Measure
   the Attitudes of Teachers Toward Computers.   Doctoral Dissertation,
   West Virginia University, 1975.   76-11,745.

5

Student Variables

Many of the variables which affect mathematics learning reside within the student himself. A good deal of effort has gone into the study of some of these variables.

No standard way of categorizing these variables has been adopted, but the categorization which I have used in this chapter is pretty much in accord with the terminologies in the literature.

Before turning to brief surveys of individual variables, four reviews by Aiken, (1), (2), (3), and (4), need to be mentioned. These reviews, which are selective rather than exhaustive, deal with a number of different categories of the variables discussed below. Together these reviews provide a helpful introduction to the material dealt with in this chapter.

It should also be mentioned that the overlap between my set of references and the ones in the bibliographies contained in these reviews is quite small, due in part to the fact that the reviews were prepared in the very early seventies.

Affective Variables

The attitudes and feelings which students have about mathematics have been classified and studied under a number of different headings.

Anxiety [27]

In some students, mathematics arouses a certain amount of anxiety, and the degree to which this happens varies from student to student. Most of these studies investigated the relationship between anxiety and mathematics achievement. The findings have been quite consistent: the correlation between mathematics anxiety and mathematics achievement is negative. The higher the degree of anxiety, the less the achievement.

Unfortunately, these studies merely establish a correlation. No direction of causality can be derived from them. However, five of the studies investigated procedures intended to reduce mathematics anxiety. If such procedures can be perfected, then it will be possible to carry out the experiments to determine whether poor achievement causes anxiety or vice versa. Work on this problem (and similar problems for other affective variables) should be given high priority.

Mathematics Attitudes [93]

About half these studies merely measured student attitudes toward mathematics. As in the case of anxiety, attitudes do vary from student to student. Two of the findings are of interest. First, despite a widespread belief, mathematics (or arithmetic) is not the most strongly disliked school subject. In fact, the average student attitude towards mathematics is a neutral one. Thus, when elementary school children are asked to order their school subjects (mathematics, English, science, social studies, etc.) in decreasing order of liking, mathematics, on the average, ends up in the middle.

The second finding, which shows up very clearly in the NLSMA data, is that while attitudes towards mathematics improve slightly between fourth and sixth grades, once students reach secondary school their attitudes towards mathematics decline slowly over time. This decline has since been noted in a significant number of other studies. No explanation of this decline has been found.

It should be mentioned that a variety of different ways of measuring student attitudes towards mathematics have been used. A careful study of the interrelationships between these different measures would be useful. Even more useful would be agreement on a few standard measuring instruments, for this would make it easier to interrelate the results of different investigators, something which has happened very little in the past.

Also, as Aiken and some others have pointed out, it might be helpful to replace the global measures of mathematics attitudes which have been primarily used so far with measures of more specific attitudes toward more specific aspects of mathematics. NLSMA did something along this line by developing tests of different kinds of attitudes, but further development is needed. And measures of attitudes toward specific aspects of mathematics are conspicuous by their rarity. Separate attitude measures for each of the cells of the NLSMA matrix shown in Chapter Two would be very useful.

The remaining studies are divided about equally between investigations of the relationship between mathematics attitudes and mathematics achievement and investigations of procedures intended to improve student attitudes. As in the case of anxiety, there is a significant positive correlation between attitudes and achievement. Positive attitudes go with greater achievement. The relationship, however, is not as strong as many seem to believe, and it does not show up in every study. Also, as in the case of anxiety, we do not know in which direction causality runs between attitudes and achievement.

There is hope that this matter can eventually be settled, since there has been some work on procedures for modifying student attitudes, and the probability of success on this seems to be at least greater than zero. I am sure I am not alone in believing that it would be an unsatisfactory state of affairs if our students were to wind up disliking mathematics or having mathematics create anxiety in them, even if their mathematics achievement were not affected. Research along these lines is to be encouraged.

Finally, a very few studies compared teacher attitudes with student attitudes. Again contrary to general belief, they seem not to be related. And, of course, it was pointed out in Chapter Three that teacher attitudes have little effect on student achievement. However, the number of studies here is too small to allow any firm conclusion, and further research is very much called for.

Motivation [18]

The situation with this variable is much as it is for attitudes towards mathematics. More than half the studies relate motivation to mathematics achievement. The correlation is significant and positive but not strong. The rest of the studies investigate the levels of motivation in different sets of students or investigate procedures which might change motivation. As in the case of attitudes, further research could have a good payoff.

Personality [29]

A number of variables have been studied that can be placed under the heading "personality." However, none of the relationships with mathematics achievement seem very strong. In addition, the number of personality variables which have been looked at is relatively large, and the relationships between them are not clear.

It is difficult to decide what use can be made of these studies or whether further research along these lines would be rewarding.

School Attitudes [8]

Except that they have been less often studied, student attitudes towards school are much like their attitudes towards mathematics. There is some correlation between school attitudes and achievement, but which causes which is unknown. The studies to date are so scattered and unrelated that no clear patterns emerge.

Self-Concept [38]

A student's opinion of his own ability in mathematics is an attitude of particular interest both to educators and to laymen. Most of the studies undertaken so far have been concerned with the relationships between mathematics self-concept and mathematics achievement. The usual finding is that the correlation is significant and positive, but not overwhelmingly large. As in the case of other attitudes, it is not clear which causes which.

A more detailed understanding of mathematics self-concept and its relationships to achievement would be useful. A standardization of self-concept measures would contribute to such studies.

A handful of studies were concerned with efforts to change students' self-concepts. Not enough has been done along this line to allow any firm conclusions.

Test Anxiety [16]

Anxiety aroused specifically by mathematics tests, rather than by mathematics in general, has received some attention. The findings have been quite similar to those for general anxiety. In particular, a sex difference seems likely, with girls showing more test anxiety than boys.

Cognitive Variables

Psychologists have identified and studied several dozens of different dimensions along which individuals can vary as they carry out cognitive activities. We use "cognitive factor" as a generic term for these cognitive dimensions. We give three examples.

i) When faced with a question or a problem, some individuals react quickly with an impulsive response. Others, however, seem to take time to reflect on the matter before responding. The available evidence indicates that reflective students do better in mathematics than do impulsive ones.

ii) Another dimension is measured by a test in which each item presents four geometric designs followed by a complex geometric figure. The subject is asked to decide which of the designs is embedded in the figure. The designs are the same for all items, while the figures vary from item to item. The score on this test is the length of time needed to answer all items correctly. Subjects with low scores are called "field independent," and the evidence is that they do better in mathematics than do field dependent students.

iii) Individuals vary considerably in their ability to visualize and mentally manipulate geometric objects. (This dimension, incidentally, is not highly correlated with the preceding one.) It would seem that students high in spatial visualization would do better in geometry than students low on this dimension, but the evidence for this is quite weak.

These three and many of the other cognitive factors which psychologists have identified have been studied to see how they affect the learning of mathematics. Readers not familiar with this part of the psychological literature will find a recent review by Ekstrom (5) helpful.

In what follows I separate out certain cognitive factors, or combinations of them, for individual mention, since they have each received a good deal of attention. The rest I lump together into one miscellaneous container.

IQ [25]

Most intelligence tests are made up of a mixture of measures of separate cognitive factors. The mixture varies from test to test, which explains why the administration to two different IQ tests to the same student often results in two different IQ's for that student.

Few of the references I have collected under this heading are devoted merely to demonstrating that there is a significant positive correlation between mathematics achievement and IQ. That fact is too well established to need any further confirmation. What is generally overlooked, however, is that the degree of correlation is only moderate and that IQ accounts for only a modest amount of the variance in mathematics achievement scores.

A number of interesting questions related to IQ have been studied. For example, how do students high on verbal IQ but low on quantitative IQ compare with students with the opposite IQ profile? How is IQ related to creativity? Do theories of intelligence such as those of Piaget (6) or Guilford (7) provide more useful (for mathematics education) IQ measures than the customary ones? So far, only fragmentary answers to such questions are available.

Logical Thinking [14]

The domain of this variable is quite fuzzy, and a number of different phrases--"deductive reasoning," "logical reasoning," "thinking skill," "creative thinking," "critical thinking"--are used in describing a variety of studies of ability which seem to be loosely related. Only a few of these studies focus on mathematics.

A review by Jansson (8) of the recent literature on deductive reasoning is helpful.

While the scanty literature provides us with little guidance as to how we can help our students improve their reasoning and thinking skills, there is nothing in this literature to diminish our hopes that some day we might succeed in this task. This topic seems to deserve a substantial, well-coordinated research effort.

Mathematics Ability [118]

Three different kinds of studies fall under this general heading. First are those that ask the question: "How is mathematical ability defined, and how do we measure it?" The usual procedure in these studies is to determine, by administering large batteries of tests, which cognitive factors (including IQ) best predict success in mathematics and to attempt to compute that particular linear combination of these cognitive factors that does the best job of prediction.

The second kind of study accepts a particular definition of mathematical ability, usually based on a standardized test, and then studies the differences between students at different levels of this ability.

The third kind of study combines these two approaches but restricts attention to a particular topic in mathematics rather than to mathematics in general.

One great virtue of these studies is that they make it very clear that mathematical ability is multivariate and that no single number, such as an IQ, is a useful predictor of mathematics achievement.

In general, however, these studies have not proved very helpful. Batteries of tests purporting to measure mathematical ability usually leave unaccounted for a substantial portion of the variance in achievement scores. If there are indeed specific cognitive factors which determine general mathematical ability or even ability to learn and do specific mathematical topics, we do not yet know what they are.

Memory [10]

Psychologists distinguish between short term and long term memory. The first can be measured, for example, by the number of correctly recalled digits immediately after a sequence of random digits has been presented, either orally or visually. This number seems to cluster around seven, but shows quite a bit of variability from person to person.

An example of long term memory is one's ability to remember one's area code and telephone number.

It seems likely that both kinds of memory are important for mathematics learning, but little research seems to have been done on this topic. It is hard to guess, from the studies which have been done, what payoff there would be from further studies.

Miscellaneous Factors [75]

Among the studies placed in this category, spatial visualization is the most frequent cognitive factor investigated, and the findings for this factor are typical of those for most factors: there is a small but significant correlation between measures of spatial visualization and mathematics achievement.

Thus, for example, in my study, NLSMA No. 27, of the predictors of mathematics achievement, a number of cognitive factors did show some ability to predict mathematics achievement, but they were almost invariably less effective as predictors than was previous mathematics achievement.

Reading [22]

It would be surprising if there were no correlation between reading ability and mathematics ability. On the basis of the relatively small number of studies of this and related questions, it appears that the correlation is indeed positive, but not very large. However, since we know that instruction can improve reading ability, more information on the relationships between reading and mathematics is clearly called for.

Knowledge

There have been a substantial number of studies which merely asked how much various students knew about various mathematical topics. In these studies, no measures were taken of curriculum or teaching variables. Those studies in which such measures were obtained would have been listed elsewhere.

These studies, even though they provide nothing more than the status of knowledge about certain mathematical topics, do play a useful role. Some of the topics, such as probability, geometry, and logic, have been measured before the students have actually met them in their studies. These surveys then provide information about the understanding of a topic, usually quite intuitive, which students can be expected to have at the beginning of a systematic study of the topic. They can therefore be quite useful to the curriculum developer.

Even for topics which the students had studied in school, status reports can be useful. Thus, for those developing subject matter courses for prospective elementary school teachers, it is helpful to know how much the students have retained, if in fact they were ever taught them, of the mathematical concepts they should have met in the elementary school.

In addition, status reports at various grade levels can sometimes pinpoint common failure patterns that might otherwise go unnoticed or can bring to light gross deficiencies in the curriculum of a city, state, or country. A case in point is the survey of mathematical understanding carried out by the National Assessment of Educational Progress: (9) and (10).

There have been numerous different mathematical topics which have been surveyed. I comment below on seven of these which seem worthy of separate mention, and point out the existence of others which I have placed under the heading "Miscellaneous."

Preschool Mathematics Knowledge [17]

A number of these studies asked what students knew at entrance to kindergarten and others at entrance to grade one. Most asked about number concepts, but in some cases attention was paid to geometry and to arithmetic vocabulary.

It is clear that some students, even at the beginning of kindergarten, possess some intuitive arithmetic ideas. However, a careful review of all these studies would be needed before any generalizations could be stated.

Arithmetic [57]

A number of different emphases are found in these studies. Some compared number knowledge in different countries or at different points in time. Others investigated the arithmetic understanding of prospective elementary school teachers. Still others looked at the sequential development of number concepts in young children. The influence of Piaget is strong in these last studies. A review by Bernbaum (11) is helpful.

Finally, there are numerous status reports for various grade levels, some of which concentrate on typical errors which students make.

Geometry [23]

Most of the studies under this heading dealt with younger children and were concered with the development, over time, of geometric concepts.

Logic [31]

The subjects in these studies included adolescents and college students as well as young children. Even the latter showed some intuitive grasp of logic ideas. A substantial proportion of the studies were interested in the developmental aspects of logic comprehension.

Probability [13]

Most of these studies were done with elementary school students and tried to investigate the development over time of probability concepts.

It seems clear that even young children, without having had any instruction on probability, acquire some of the basic concepts quite early. The indication is that it should be quite feasible to teach some of these concepts in the elementary school. (SMSG had already demonstrated this by means of the elementary school probability units which it developed.)

Ratio and Proportion [14]

These topics are unlike the others above, since they are normally taught in the late elementary or early secondary program. These studies indicate that younger children find it difficult to apply these concepts to real-life situations.

Measurement [11]

In addition to measurement of length, area, volume, and weight, time concepts have also been studied. Almost all the studies involved only elementary school students. A quick glance at the reports of these studies does not reveal anything startling. A detailed review of them would be worthwhile.

Miscellaneous [32]

A wide variety of other topics (e.g., estimation, concept of function, graphs, algebra) have been investigated, but in none of these cases have there been enough studies to allow any generalizations to emerge.

Nonintellective Variables

There are student variables other than the affective and cognitive ones discussed above which might affect the learning of mathematics. I place these variables in four separate sets.

Ethnicity [34]

There have been many studies in which the student's ethnic affiliation was one of the variables. The number above is substantially smaller than the total number of such studies, since I have included here only those in which, besides the ethnic variable, only mathematics achievement was measured without looking for its relationship to, or correlation with still other variables.

The results of these studies are well known. Black, Chicano, and American Indian students do less well in mathematics than Anglo students. There is some slight evidence, though, that the more Chicanos and American Indians depart from their native cultures and move toward the middle class culture of the United States, the better their achievement in mathematics. More information on this would be useful.

Physical Variables [36]

Included under this heading are a number of studies on the effects of age at entrance to school. The results of these studies were mixed. Other studies looked at birth order, at physical fitness, etc. The findings provide no real surprises and no clear suggestion for the improvement of mathematics education.

Socioeconomic Status [26]

Studies of the relationship between this variable and mathematics achievement provide no surprises. Lower SES students do less well in mathematics.

Sex [58]

There have been enough studies to make it clear that girls do better than boys on computation, while boys do better than girls on higher level mathematical tasks, at least from the upper elementary grades through the first year or two of high school. In addition, girls elect fewer mathematics courses in high school than do boys. This severely limits girls in their choices of college majors and hence restricts their possible career choices.

Although the question is still open, it seems likely that these sex differences are sociological rather than genetic in nature. This question is in need of further research.

A useful review of the literature is provided by Fennema (12).

Predictors of Achievement [55]

There have been numerous attempts to discover what kinds of information we should have about a student at one point in time in order to predict that student's mathematics achievement at a later point in time. Many different kinds of variables have been studied, cognitive, affective, and nonintellective. Some of the studies merely concentrate on one particular variable to see if it has any predictive power. (It generally turns out that it does have some predictive power, but not of any great magnitude.)

More often, the studies have used a variety of different measures and have employed statistical procedures to sort out the most effective combination of them. Such studies have been carried out for many different grade levels, ranging from primary school to college. But the

particular prediction which has received the most attention is that of
achievement in the beginning high school algebra course. The decision,
at the end of a student's eighth grade, as to whether he should enroll
in the beginning algebra course in grade nine or should be placed in a
"general mathematics" or some other less demanding course, or the decision at the end of a student's seventh grade as to whether he should
start algebra early, has to be faced in every school system. Erroneous
decisions, in either direction, are bad for the student, so there have
been many attempts to provide relevant information to the decision makers.

I have reviewed most of these studies (13). It turns out that the
best predictors of success in beginning algebra are measures of the student's previous success in mathematics, as measured by his grades in
mathematics courses or by the opinions of his mathematics teachers.
General intellectual ability, as measured by IQ, and reading ability seem
to have little value as predictors of algebra success.

Probably the most massive investigation of predictors of mathematics
achievement ever carried out was one which I conducted using NLSMA data
and reported in NLSMA Report No. 27. Each of the mathematics scales
used in NLSMA from the end of the first year to the end of the last year
of the study was investigated for each of the three student populations.
For each of these scales all available information about the students was
investigated: previous mathematics achievement, cognitive and affective
variables, nonintellective variables such as SES, and everything that
was known about the teachers. Stepwise regression was used to determine
those variables which contributed significantly to the prediction of
scores on each scale.

From this extensive investigation two generalizations emerge very
clearly:

i) The best predictor of mathematics achievement is previous
mathematics achievement. Only rarely does any other cognitive variable
contribute significantly to prediction. Almost never do affective, nonintellective, or teacher variables add anything to the predictive power
of previous mathematics achievement measures.

(Also, as mentioned earlier, the best predictor of computational
achievement is previous computational achievement, and the best predictors
of achievement at higher cognitive levels are previous mathematics
achievement at higher cognitive levels.)

ii) Despite the very large numbers of variables examined as potential predictors, none of the achievement variables were predicted very accurately. The stepwise procedure used provided, for each scale, the amount of the variation in scores on that scale that was due to the significant predictors as a percentage of the total variation. The following table summarizes the results:

Table 5.1

Percentage of Variance Explained by Significant Predictors

Population	Number of Scales	Median	Range
X	78	43%	10% – 69%
Y	76	40%	6% – 68%
Z	16*	37.5%	22% – 66%

*For the Z-Population, no analysis was made for the scales administered during the third year, since the variety of courses taken was so large that the number of students per course was too small. And, of course, the Z-Population graduated from high school at the end of the third year and was given no more NLSMA tests.

The other studies parallel this one as far as the amount of variance in the criterion variable which the significant predictors account for is concerned. It is normally rather low. Like human beings in general, students are not very predictable.

Study Behavior [3]

Some research has been done on the effects of such student activities as underlining in textbooks, or taking notes during lectures. Some significant findings have been reported, but I have located only three studies along these lines which involved the learning of mathematics. It would be reasonable to carry out further exploratory studies in this area at least at the college level.

This survey of student variables provokes mixed reactions. Thus, we should be pleased that so much has already been learned about the role of affective variables in mathematics education, and we should encourage further research in this area. I am sure that most readers would agree with me that positive attitudes towards mathematics, even if they did not correlate at all with achievement, would be intrinsically desirable. Efforts to find ways to improve attitudes should be encouraged.

The status of student knowledge about mathematics is well worth knowing, both as a guide for curriculum developers and as a means of monitoring current mathematics programs. However, we cannot be satisfied with the present state of affairs. Thus, for example, the National Assessment of Educational Progress does an excellent job in testing a national sample of students at various grade levels. Yet its coverage of mathematics is too thin to provide much guidance to curriculum developers or to allow for anything but a very gross monitoring of current mathematics programs. On the other hand, those investigations that have studied specific topics more deeply have been plagued by inadequate samples. There is clearly room, and need, for more extensive work in this general area.

Cognitive variables provoke another and more discouraging reaction. The reason is that these variables all seem to be quite stable over time and quite unresponsive to instruction. Therefore we must ask what good it does us to know that one student is high in a particular cognitive factor while another student is low on that factor. We observed in the previous chapter that we have, so far, been unable to adapt instructional programs to such differences. What, then, can we do?

The only answer that has been suggested so far is to select for instruction only those students who are high on those cognitive factors which facilitate achievement on the particular mathematical topic under consideration and to deny instruction to those students who are low on those factors. Such a suggestion is of course damned as being unfair and undemocratic.

On the other hand, we actually do carry out such selections to a certain extent. When children are very young, those with very low IQ's are placed in institutions for the mentally retarded and are not provided the normal school experience. Starting in the elementary school, in some cases, some of the more able students are instructed separately and are given a richer program. (This matter is taken up in the next chapter.) In junior high school, decisions are made, presumably at least in part on the basis of measures of some kinds of ability factors, as to which students will be allowed to take the beginning course in algebra. Later, decisions on admission to college, and still later, on admission to graduate school, are based in part on ability measures.

It is probably the case that neither professional educators nor the lay public in this country would approve of going much beyond the selection filters mentioned above. (This is not true in some other societies.) Consequently, an educational researcher interested in cognitive variables should start by asking whether the results of his proposed research could have any practical educational applications. (I am not suggesting an end to investigation of cognitive variables. I am still hopeful that our colleagues in psychology, despite their lack of success so far, will eventually develop some theoretical understanding of these variables which we can use for the improvement of mathematics education.)

What I have said about cognitive variables is equally true about predictors. There is no point in sifting through fourth grade information, for example, in order to locate the best predictors of fifth grade achievement if no use is going to be made of such predictors.

The nonintellective variables are still another story. We need no further studies to convince us that there are sex differences in mathematics achievement or that there are differences due to ethnic status. What we do need is a deep enough understanding of the reasons for these differences so that we can begin to do something about remedying them. I do not believe that this task can be carried out successfully by educators alone. Nor do I believe that it can be left solely to others, presumably our social science colleagues. I believe that only an intensive joint effort has any chance of success.

Bibliography

1. Aiken, L. R. Nonintellective Variables and Mathematics Achievement:
    Directions for Research. *Journal of School Psychology,* Vol. 8
    (1970) pp. 28-36.
2. Aiken, L. R. Attitudes Toward Mathematics. *Review of Educational
    Research,* Vol. 40 (1970) pp. 551-596.
3. Aiken, L. R. Intellective Variables and Mathematics Achievement:
    Directions for Research. *Journal of School Psychology,* Vol. 9
    (1971) pp. 201-212.
4. Aiken, L. R. Ability and Creativity in Mathematics. *Review of
    Educational Research,* Vol. 43 (1973) pp. 405-432.
5. Ekstrom, R. B. Cognitive Factors: Some Recent Literature. ED 080 596.
6. Flavell, J. H. *The Developmental Psychology of Jean Piaget.*
    Princeton, N.J.: D. Van Nostrand Company, 1963.
7. Guilford, J. P., and Hoepfner, R. *The Analysis of Intelligence.*
    New York: McGraw-Hill, 1971.
8. Jansson, L. C. The Development of Deductive Reasoning: A Review of
    the Literature, Preliminary Version. ED 090 034.
9. Carpenter, T. P., and others. Results and Implications of the NAEP
    Mathematics Assessment: Elementary School. *Arithmetic Teacher,*
    Vol. 22 (1975) pp. 438-450.
10. Carpenter, T. P., and others. Notes from National Assessment:
    Word Problems. *Arithmetic Teacher,* Vol. 23 (1976) pp. 389-393.
11. Bernbaum, M. Number and Concept Development: An Abstract Biblio-
    graphy. ED 057 921.
12. Fennema, E. Mathematics Learning and the Sexes: A Review.
    Journal for Research in Mathematics Education, Vol. 5 (1974)
    pp. 126-139.
13. Begle, E. G. Predicting Success in Beginning Algebra: A Review of
    the Empirical Literature. SMESG Working Paper No. 18. ED 121 606.

Aiken, L. R. Personality Correlates of Attitude Toward Mathematics. *Journal of Educational Research,* Vol. 56 (1963) pp. 474–480.

Abrego, M. B. Children's Attitudes Toward Arithmetic. *Arithmetic Teacher,* Vol. 13 (1966) pp. 206–208.

Bachman, A. M. The Relationship Between a Seventh-Grade Pupil's Academic Self--Concept and Achievement in Mathematics. *Journal for Research in Mathematics Education,* Vol. 1 (1970) pp. 173–179.

Baer, C. J. The School Progress and Adjustment of Underage and Overage Students. *Journal of Educational Psychology,* Vol. 49 (1958) pp. 17–19.

Barton, K., Dielman, T. E., and Cattell, R. B. Personality and IQ Measures as Predictors of School Achievement. *Journal of Educational Psychology,* Vol. 63 (1972) pp. 398–404.

Battle, E. S. Motivational Determinants of Academic Competence. *Journal of Personality and Social Psychology,* Vol. 4 (1966) pp. 634–642.

Chansky, N. M. Anxiety, Intelligence and Achievement in Algebra. *Journal of Educational Research,* Vol. 60 (1966) pp. 90–91.

Cox, F. N. Test Anxiety and Achievement Behavior Systems Related to Examination Performance in Children. *Child Development,* Vol. 35 (1964) pp. 909–915.

Dwyer, C. A. Influence of Children's Sex Role Standards on Reading and Arithmetic Achievement. *Journal of Education Psychology,* Vol. 66 (1974) pp. 811–816.

Epps, E. G., et al. Effect of Race of Comparison Referent and Motives on Negro Cognitive Performance. *Journal of Educational Psychology,* Vol. 62 (1971) pp. 201–208.

Feij, J. A. Field Independence, Impulsiveness, High School Training, and Academic Achievement. *Journal of Educational Psychology,* Vol. 68 (1976) pp. 793–799.

Fitzgerald, W. M. On the Learning of Mathematics by Children. *Mathematics Teacher,* Vol. 56 (1963) pp. 517–521.

Guilford, J. P., Hoepfner, R., and Petersen, H. Predicting Achievement in Ninth-Grade Mathematics from Measures of Intellectual-Aptitude Factors. *Educational and Psychological Measurement,* Vol. 25 (1965) pp. 659-682.

Keough, J. J. The Relationship of Socio-Economic Factors and Achievement in Arithmetic. *Arithmetic Teacher,* Vol. 7 (1960) pp. 231-237.

Knifong, J. D., and Holtan, B. D. A Search for Reading Difficulties Among Erred Word Problems. *Journal for Research in Mathematics Education,* Vol. 8 (1977) pp. 227-230.

Majoribanks, K. School Attitudes, Cognitive Ability, and Academic Achievement. *Journal of Educational Psychology,* Vol. 68 (1976) pp. 653-660.

Rea, R. E., and Reys, R. E. Competencies of Entering Kindergarteners in Geometry, Number, Money, and Measurement. *School Science and Mathematics,* Vol. 71 (1971) pp. 389-402.

Skemp, R. R. Reflective Intelligence and Mathematics. *British Journal of Educational Psychology,* Vol. 31 (1961) pp. 45-55.

Whimbey, A. E., and Ryan, S. F. Role of Short-Term Memory and Training in Solving Reasoning Problems Mentally. *Journal of Educational Psychology,* Vol. 60 (1969) pp. 361-364.

6

The Environment

In addition to teachers, curriculum, and the student's own inherited
and acquired characteristics, the environment in which he is embedded can
affect his learning of mathematics. A number of different aspects of the
environment have been studied.

Ability Grouping [76]

This term is used when students are assigned to classes in such a
way that the variation in ability within any one class is smaller than
it would have been if the students had been assigned randomly. A synonym
is "homogeneous grouping."

This practice has resulted in a considerable amount of discussion,
much of it quite emotional, both pro and con. Those in favor make such
claims as the following: teachers can work more effectively in classes
with a restricted range of student abilities; the brighter students are
held back by the presence of the less able students in their classes;
or the less able students are discouraged by the presence of the able
students in their classes.

Those opposed to ability grouping claim that the less able students
need the inspiration which abler students in their classes can provide,
that they feel demeaned by being segregated, and that the brigher students are likely to develop undesirable feelings of superiority if segregated away from the ordinary students.

Fortunately, there has been a good deal of empirical study of the effects of ability grouping. I have reviewed (1) most of these studies and can report that most of the passion expressed in the discussions mentioned above was wasted. The evidence is quite clear that the most able students should be grouped together, separate from the rest of the student population. When this is done, these high ability students learn more mathematics than they would otherwise and do not develop any undesirable attitudes. At the same time, the remaining students do just as well on both cognitive and affective variables without the very able students in class with them as they would if the very able students were in their classes. On the other hand, for all but the most able students, it seems to make little difference whether they are grouped homogeneously or not.

Another finding which emerges from these studies is that when grouping is to be used, it should be done on the basis of previous mathematics achievement rather than on the basis of general ability as measured by, for example, overall grade point average or IQ.

Acceleration [54]

The term is, of course, a misnomer. What is referred to is the practice of having high ability students proceed through the curriculum at a more rapid rate than the average students follow. I have also found a few studies in which low ability students were allowed to proceed at a slower pace. Logically, this topic might have been postponed to the next chapter, on teaching methods, but it is more convenient to put it here since "acceleration" is used primarily with homogeneously grouped high ability students.

I have reviewed (2) most of the empirical studies which fall under this heading. The findings are less definitive than those for ability grouping. Thus, it is not clear whether the talented students should be allowed to continue in the regular curriculum throughout the school year, learning more of that content than the average students. An alternative wou be to have the talented students cover only as much of the regular curriculum as the average students but to use the time saved by moving at a more rapid pace in studying "enrichment" topics, i.e., topics which are interesting but off to the side. The results of the studies of this question are mixed.

Thus, while it seems likely that acceleration is often appropriate for talented students, still other questions remain. Two such questions are: At what grade level should acceleration start? Should low ability students be permitted to proceed at a reduced pace? This is clearly a topic which deserves considerably more research.

Class Climate [17]

This ambiguous phrase is used in connection with a number of different social and psychological aspects of the classroom, and a number of different measuring devices have been developed. Whatever it is, classroom climate does seem to have a small but significant effect on mathematics learning. A closer look at the studies already carried out, including perhaps some of those involving other subject matter areas, would at least clarify the various variables which have been investigated and might provide some useful suggestions for further research.

Class and School Organization [84]

A number of different organizational patterns have been studied, including "open" vs. traditional classrooms, departmentalized vs. self-contained classrooms, graded vs. ungraded systems, and single-sex vs. combined-sex classrooms. For most of these patterns, there have been some significant differences in favor of each end of the dimension, but there have also been many cases in which no significant differences appeared.

A careful review of all these studies seems needed. A review of research on ungraded classrooms was prepared by the University of Georgia (3) but is now ten years old. A more recent review by Goodlet (4) of the studies conducted during the sixties concludes that, largely because of methodological problems, it is not yet possible to reach a decision on ungraded classrooms.

It is generally believed that teachers are more effective and that students learn more in small classes than in large ones. The research literature, however, casts severe doubts on this belief, as is indicated in several reviews of the empirical studies. There have been, as yet, no reviews devoted exclusively to the size of mathematics classes, but a cursory survey of those studies which I have been able to locate supports the view that what is true in general is also true for mathematics classes in particular. A very few of these studies report an advantage for small classes, but an equal number are on the opposite side. Most of the studies indicate no significant differences in mathematics learning which can be attributed to class size. A careful review of these studies is needed.

Some of the studies included under this heading investigated the advantages of having students work, in the classroom, in small groups of two or three. There were no strong findings, but a more careful look at these studies would also be worthwhile.

Ethnic Environment [14]

What does it mean to a black student studying mathematics if the percentage of white students in his class is 60 rather than 6? The results of the largest study along this line are to be found in the Coleman Report (5). Unfortunately, this report is not relevant to my interests because the study looked only at verbal, not mathematics, achievement.

There have been some other studies, however, that have looked at this general question, and, in particular, have investigated mathematics achievement in schools before and after they were desegregated. Although I have not had a chance to study these in detail, there are hints that the effects of desegregation on mathematics achievement may not be as strong as one might wish and also that they might be different from the effects of verbal achievement. A review of many of these studies by Bradley and Bradley (6) found that almost all of them suffered from methodological weaknesses.

It is clear that we know much less than we ought to about this important variable.

Family Background [37]

A number of different family variables have been investigated. These include father absence, child-rearing practices, parental behavior and attitudes, home environment (as measured by specially designed questionnaires), and family mobility. In many of these studies significant achievement differences appear, but for no single family variable is there enough information to suggest a strong pattern of influence on mathematics achievement.

Further research along these lines should certainly wait until these studies have been carefully reviewed.

Physical Variables [19]

Some physical characteristics of the classroom have been investigated. A small number of studies suggest that temperatures over 80 (27°C) result in impaired performance on mathematical tasks, such as addition exercises. However, it should be noted that performing such a mathematical task is not the same as learning a new mathematical idea. It is especially important to keep this distinction in mind in connection with studies in which a physical variable is found to have no significant effect.

Thus, a small number of studies found consistently that a moderate amount of background noise has no effect on performance on a mathematical task. Similarly, atmospheric pressure seems to have no effect on mathematical performance. One study found that the number of square feet of classroom space per student was not related to mathematical performance.

It seems clear that much more study of these variables is needed. The financial implications, if the above findings were to be replicated with higher cognitive level criterion measures, could be enormous. Thus, air-conditioning might become imperative in many parts of the country. On the other hand, the need for noise-proofing, or for spacious mathematics classrooms, might be drastically reduced with great consequent savings.

My reactions to the research sketched in this chapter are similar to my reactions to the topics discussed in the preceding chapter. They are mixed. Some topics--in particular, Ability Grouping--seem to call for little or no more research, at least at present. Others--including Acceleration, Ethnic Environment, and Physical Variables--on the other hand, seem to me to be worth a substantial research effort.

For the remaining topics, I suggest a careful review of the relevant research literature, including some of that done in other subject matter areas, before reaching any decisions on the need for further research in these areas.

Bibliography

1. Begle, E. G. Ability Grouping for Mathematics Instruction: A Review of the Empirical Literature. SMESG Working Paper No. 17 (1975). ED 116 938

2. Begle, E. G. Acceleration for Students Talented in Mathematics. SMESG Working Paper No. 19 (1975). ED 121 607

3. University of Georgia, Research and Development Center in Education. Abstracts of Research Pertaining to Ungraded vs. Self-Contained Classroom Organization in the Elementary School. ED 019 129

4. Goodlet, G. R. Nongrading and Achievement: A Review. *Alberta Journal of Educational Research,* Vol. 18 (1972) pp. 237-242.

5. Coleman, J. S., and others. Equality of Educational Opportunity. Office of Education (1966) U. S. Government Printing Office.

6. Bradley, L. A., and Bradley, G. W. The Academic Achievement of Black Students in Desegregated Schools: A Critical Review. *Review of Educational Research,* Vol. 27 (1977) pp. 399-449.

Illustrative Research Reports

Carlsmith, L. Effect of Early Father Absence on Scholastic Aptitude. *Harvard Education Review*, Vol. 34 (1964) pp. 3-21.

Guggenheim, F. Classroom Climate and the Learning of Mathematics. *Arithmetic Teacher*, Vol. 7 (1961) pp. 363-367.

Halliwell, J. W. A Comparison of Pupil Achievement in Graded and Non-Graded Primary Classrooms. *Journal of Experimental Education*, Vol. 32 (1963) pp. 59-64.

Ivey, J. O. Computation Skills: Results of Acceleration. *Arithmetic Teacher*, Vol. 12 (1965) pp. 39-42.

Johnson, M., and Scriven, E. Class Size and Achievement Gains in Seventh- and Eighth-Grade English and Mathematics. *School Review*, Vol. 75 (1967) pp. 300-310.

Kassinove, H. Effects of Meaningful Auditory Stimulation on Children's Scholastic Performance. *Journal of Educational Psychology*, Vol. 63 (1972) pp. 526-530.

Rea, R. E., and Reys, R. E. Mathematical Competencies of Negro and Non-Negro Children Entering School. *Journal of Negro Education*, Vol. 40 (1971) pp. 12-16.

Schrank, M. W. R. A Comparison of Academic Achievement in Mathematics of Ability Grouped Versus Randomly Grouped Students. *Journal of Educational Research*, Vol 62 (1968) pp. 126-129.

7

Instructional Variables

What students learn about mathematics depends not only on the characteristics of their teacher, the curriculum which they follow, their own characteristics, and their environment but also on the way in which they are taught. A large number of teaching variables, dimensions along which the instructional process can vary, have been studied. In fact, more experimental investigations of teaching variables have been conducted than has been the case for any of the other major headings we review in this survey.

Calculators [52]

The idea that calculators might be helpful teaching aids must have crossed the minds of many mathematics educators. Fortunately, a substantial number of them thought it worthwhile to check the idea empirically. Two useful annotated bibliographies have been prepared: (1) and (2).

Desk Calculators

In some cases desk calculators led to better achievement, but in about twice as many cases there was no noticeable effect on achievement. In any case, due to lower costs, wider availability, and greater capability of electronic calculators, further study of desk calculators seems not to be needed.

Hand-held Calculators

In about half of the studies, the use of hand-held calculators re-sulted in improvement in student achievement. However, in half of these favorable cases, the improvement was demonstrated only for computational skills.

In almost all these studies, the calculator was used merely as a supplement to a regular course. We have yet to see the results of evaluation of instructional programs which explicitly make use of the special capabilities of calculators.

Classroom Behavior [69]

The behavior in the classroom of teachers, and to some extent of students, has been studied intensively. Those particularly interested in this topic will find the review of Dunkin and Biddle (3) of interest.

A review more useful to us is that of Rosenshine (4) and the summary of it by Rosenshine and Furst (5). Three comments are needed. First of all, we referred to (4) in our discussion of teacher characteristics in Chapter Three, and we refer to it again here. The reason for the dual reference is this. Some teacher behavior defines teacher characteristics, such as "business-like" or "warmth," which seem to be quite stable and not easy to modify. These were covered in Chapter Three. Other teacher behavior (e.g., use of criticism, use of student ideas) seems much easier for a teacher to control or for a teacher training program to influence. This is covered in this chapter.

Secondly, teacher classroom behavior has been assigned to a number of distinct categories, but only a limited number of studies within each category have included information on student achievement. Of those few that do, only a very limited minority refer to achievement in mathematics.

Thirdly, confusing matters somewhat, within many of the categories a variety of measuring devices have been applied to teacher behavior, and the relationships between these devices are generally not known. Thus the definitions of these categories are somewhat fuzzy.

Also, it must be kept in mind that all the findings, as far as mathematics achievement is concerned, are correlational. No true experiments are reported.

It seems most useful,therefore, to summarize Rosenshine's findings about the effects of teacher behavior on student achievement in general, on the grounds that these findings, even if few or none of them are specific to mathematics achievement, may nevertheless provide fruitful suggestions for future research.

Rosenshine and Furst, in (5), list five teacher behavior variables that have strong support from empirical studies. These are:

clarity

variability

enthusiasm

task orientation/businesslike behavior

student opportunity to learn.

(For characterizations of these variables, the reader should consult (4) or (5). The "student opportunity to learn" variable seems to me to be a curriculum variable. It is mentioned explicitly in only one study.)

They list six more variables which seem to be less strong, but which have some overall experimental support:

use of student ideas and/or teacher indirectness

use of criticism

use of structuring comments

use of multiple levels of discourse

probing

perceived difficulty of course.

(The last seems to be either a curriculum or a student variable.)

Note that "businesslike behavior," "warmth," and "clarity/structuring" were considered in Chapter Three as identifying stable teacher characteristics.

As Rosenshine and Furst point out, "the above list of the strongest findings may appear to represent mere educational platitudes. Their value can be appreciated, however, only when they are compared to the behavioral characteristics, equally virtuous and obvious, which have *not* shown significant or consistent relationships with achievement *to date:*

non-verbal approval

praise

warmth

ratio of indirect behavior to direct behavior

flexibility

teacher talk

student talk

student participation

number of teacher/student interactions

student absence

teacher absence

teacher time spent on class participation

teacher experience

teacher knowledge of subject area." (p. 55)

(Note that the opinion of the importance of "warmth" deteriorated between (4) and (5).)

When we review these reports to see which ones are specifically relevant to mathematics education, the picture changes somewhat. First, the number of such studies under any one heading is too small to allow us to form any firm opinions about the relative importance of Rosenshine's categories of teacher behavior. However, if I were forced to rank order them in terms of their relationships to mathematics achievement, on the basis of the studies listed by Rosenshine, then I would make some changes in the Rosenshine-Furst classification.

Structuring, joined with clarity, would be a strong variable, as would businesslike behavior and enthusiasm, both of which I consider to be teacher characteristics. Opportunity to learn would also, of course, be a strong variable, and this would be joined by use of criticism.

Warmth would move up to the position of a moderately important variable, but indirect teacher behavior would now be considered to be unimportant, as would probing.

Two recent reviews cast serious doubts on Rosenshine's findings. Heath and Nielson (6) report that most of the studies on teacher behavior have serious methodological problems and that most of the positive findings sketched above are far from being established. Shavelson and

Dempsey-Atwood (7) asked whether teacher behavior is stable over time and report that in most cases where multiple measurements have been taken, the behavior is not stable.

On the basis of this evidence, I would conclude that further study of teacher behavior is not warranted. It is very time consuming to collect data on classroom behavior, as a second glance at (3) will confirm. To overcome the weaknesses pointed out in (6) will require even more time and effort. The payoff so far has been miniscule, and there is no reason to expect much more in the future. Our time and efforts could be employed more usefully in other directions.

However, the evidence which has been reviewed in references (3) through (7) is far from providing the whole story. I have collected about 50 additional references to studies of the effects of teacher behavior on student achievement. Most of these appeared after the publication of (3) and (4). A hasty perusal of these studies suggests that they should be given the same careful review that Rosenshine used before we abandon completely the study of teacher behavior.

Computer Assisted Instruction [62]

During the early sixties there was much investigation of various formats and designs of programmed instructional materials. It was often pointed out then that the use of a computer would make it easier to adapt programs to individual student differences by allowing the decision as to which frame the student should see next to be based on that student's performance on all the previous frames of the program. Following up on these suggestions, a number of computer-assisted instructional (CAI) programs were constructed and evaluated.

In a general review of alternatives to traditional instruction, Jamison, Suppes, and Wells (8) include a discussion of research on CAI in general, including not only mathematics but also several other subject matter areas.

A little less than one third of the studies I have located were concerned either with technical variations in programs (e.g., rate or amount of feedback) or with the effects of CAI programs on non-cognitive student behavior, principally attitudes. The remaining studies were split about evenly between those which were concerned only with student achievement in computation and those in which higher cognitive levels of student mathematical performance were involved.

In general, the findings in this general area have been positive, although the percentage of studies finding no significant differences has been larger among the studies involving higher cognitive level objectives than among those concerned only with computation.

CAI also seems to have no deleterious effects on student attitudes, although one study suggests that in primary school classrooms, heavy use of CAI may result in reduced interpersonal skills.

This seems to be another case in which further research would be reasonable. The major question mark results from the fact that computers are not as widely available as we might like and that they are expensive. But both these variables have been changing rapidly in recent years, so, at the very least, mathematics educators should keep a careful eye on this area.

Computers [29]

As was the case for calculators, computers have been looked at as possible mathematics teaching aids. A moderate amount of experimentation has been carried out. A useful bibliography has been compiled by Suydam (9).

Most of the experimental work has been done with secondary school students, but a few used elementary school students. In some cases students had access to a computer and could use it in various ways, ranging from checking solutions to problems to exploring open-ended problem situations. In other cases students merely wrote programs or constructed flow charts without actually making use of a computer.

The results of these studies have not been spectacular. In almost every case of a significant difference, the students using a computer orientation performed better than the control students. However, about the same number of studies reported no significant differences. Nevertheless, this is clearly an area about which we need much more information, and a good deal more experimentation is called for.

The cost and availability of computers may be less critical here
than in the case of CAI, especially if further work demonstrates that
work with programing languages or flow charts, without access to a computer, can lead to better mathematics learning.

Concrete Materials [104]

Concrete objects are widely used in elementary school instruction,
and to a lesser extent at the secondary school level, to illustrate or
to motivate mathematical ideas and operations. A substantial amount of
research has been carried out on various questions related to the use of
concrete materials.

Some of the studies merely compared the presence with the absence
of concrete objects during instruction. Others compared the use of concrete objects with the use of pictures or line drawings of the objects.
Still others compared the manipulation of objects by students with their
passive observation of teacher manipulation of the objects.

Some commercially available concrete teaching aids are structured
in such a way as to suggest certain mathematical ideas. A substantial
number of studies have investigated the effectiveness of these aids.
A review (10) of experiences in Canada with one of the better known of
them is of interest.

The results of all these studies are not unequivocal. While there
are no reports to the effect that the use of concrete materials resulted
in poorer achievement, there have been some cases where concrete materials did not result in better achievement but did consume more instructional time. When instructional time was not measured, as was the case
in most studies, no clear pattern emerges as to when concrete objects
result in greater mathematics achievement, and this is equally the case
for the use of structured materials.

Similarly, the use of pictures or line drawings is sometimes as
effective as the use of the actual objects, but we have as yet no way of
predicting for which mathematical objects or for which students this
will be the case. And we cannot tell in advance which students need
merely observe someone else manipulate objects in order to grasp the
mathematical idea, although we know this will sometimes be the case.

Thus we know that concrete teaching aids can play an important role in mathematics instruction, but we are far from knowing when and how to use them most efficiently. Further research along these lines is obviously needed, but a careful and detailed review of what has already been done should come first.

Correspondence Courses [6]

Only a few studies were found. They made it clear that many students can learn by correspondence, but the completion rate is generally not very high.

I believe that the U. S. Armed Forces conduct a very large correspondence program, but no studies of its effectiveness were located.

Discovery [102]

Some educators believe that students should discover mathematical facts, algorithms, concepts, and principles for themselves. They claim that students will retain mathematical objects longer and be able to apply them more widely if they are discovered rather than explicated by the teacher. An eloquent statement of this position was provided by Beberman (11). But not all educators agree, and an equally eloquent statement on the other side of the issue was provided by Ausubel (12).

Interest in this question was high even before the sixties and still continues to be high, not only for mathematics learning but also for the learning of most school subjects. This interest sparked a large amount of experimentation, but the findings of these studies did little to settle the matter one way or the other. At a conference on discovery learning held in 1966 (13), it became quite clear that the experimentation to that date had resulted in a standoff--for each experiment showing discovery to be more effective than exposition there was another experiment with the opposite result. Unfortunately, the situation today, more than ten years later, is not much different. About as many studies show mixed results or else no significant differences as show significant differences. Of the latter, more show expository teaching as more effective than discovery methods, but not enough to allow us to come to a definite conclusion.

The 1966 conference spotlighted the major methodological problem
in studies of discovery learning--the lack of a precise definition.
Obviously, there is no such thing as pure discovery. If a teacher wants
a student to discover that the sum of the angles of a triangle is 180°,
he must at least indicate to the student that he is supposed to discover
something about triangles, rather than, for example, the type style
used in the textbook or the number of inches per page. In other words,
the teacher must provide some guidance. But both the amount and the kind
of guidance can be varied, and in practice they do vary quite considerably
from experiment to experiment.

Some regularities do seem to show up in the experimental findings.
When expository teaching turns out to be significantly more effective
than discovery teaching, it is usually on the basis of an achievement
test given immediately after the experimental treatment. When the
opposite result is found, it is usually on the basis of a delayed retention test or on a test of the ability to transfer what has been learned
to a different but related situation.

Just this is enough to make it clear that we need to know more than
we do about this topic. But there are other questions which have not yet
been sufficiently addressed. For example, an expository lesson generally
takes less time than a discovery lesson. What would happen if the time
saved by the expository students were devoted to overlearning the subject matter or to learning directly the transfer task?

In another direction, it is my impression that most of the experiments have been carried out with above average students. Can all students discover? If so, what does a teacher do with his bright students,
who discover quickly, while waiting for his slower students?

There are a few scattered reports to the effect that students who
do discover successfully do better the next time and discover more
rapidly. Is this true in general, and if so what are the implications
for the student who did not discover the first time?

Another question which needs answering is: Are the different
effects of discovery and expository teaching equally strong for different
kinds of mathematical objects? There is some indication that for mathematical concepts expository teaching is definitely better than discovery.

These are but a small sample of the questions which ought to be answered by further research.

Games [22]

Games have been suggested by some educators as a teaching device, and some studies of the effects of the use of games as a part of mathematics instruction have been carried out. A cursory review of these studies does not indicate any substantial instructional power for games. Nevertheless, a detailed review of all these studies should be carried out before their potentialities are dismissed entirely.

Homework [27]

The assignment of homework is a common practice of mathematics teachers, at least above the primary level. Only a few of the studies in this area have questioned the value of homework, but together they raise serious doubts. A detailed review of these studies is needed.

The majority of the studies here compare two or more different ways of assigning or grading homework. No strong findings seem to emerge from these studies.

Individualized Instruction [193]

No two students are alike. Consequently, the optimal instructional program for any student is one which has been individualized to fit his particular abilities and interests.

That statement is, of course, tautological. Unfortunately, we seem not, as yet, to be able to take advantage of it to improve mathematics instruction in any substantial way.

During the sixties and early seventies at least three large scale curriculum projects produced individualized programs at the elementary school level, and a large number of teachers and doctoral students prepared their own individualized units at both the elementary and the secondary level. For most of these, comparisons with "traditional" instructional programs were carried out.

Fortunately, we have three reviews of these empirical studies, one by Miller (14) and two by Schoen (15) and (16).

As is pointed out in these reviews, the one individualization feature that is common to all the programs in which experimentation has been done is that they allow each student to proceed at his own pace. In very few cases did the programs allow for student variation in other abilities. This is probably not surprising in view of our lack of success (see Chapter Four) in searching for aptitude-treatment interactions.

Miller reports that about half the studies he reviewed showed no significant difference between the individualized program and the traditional one. Of the remainder, more were in favor of the individualized program than the traditional one. He found the overall advantage of individualized programs, however, to be quite small. He also found that more than three quarters of his studies reported no significant differences on affective outcomes. One interesting trend which he reports is that the longer the experiment lasted, the smaller the advantage of individualization.

Schoen paints a somewhat more discouraging picture. He found that individualization does have a positive effect, especially for affective outcomes, for the primary grades, but that, especially for grades 5-8, individualization is usually deleterious.

Schoen also points out that all the individualized programs which have been tried out so far make severe demands on the teacher's time and usually require some assistance in the form of classroom aides or computerized processing of student test results, either of which imposes a financial burden on the school.

In summary, the theoretical advantages of individualization seem difficult to realize in practice. Until new, more efficient methods of classroom management are developed, it would seem that further research in this area is unlikely to have much payoff.

In-Service Training [27]

One way of attempting to improve student achievement in mathematics has been to provide special in-service training to their teachers. Evaluation of these training programs, in terms of student achievement, has produced some positive results, though none very spectacular. A detailed review of all these studies would probably be worthwhile. Without such a review, further investigations of in-service programs would not be advisable.

Knowledge of Results [35]

Many tests are administered in mathematics classes, both for diagnostic and for grading purposes. Questions can be asked as to how much information should be provided the students as to the correctness of their responses to the test items. Also, it can be asked whether the length of time between taking the test and receiving the feedback information has any effect on the student's achievement.

A number of studies have looked into these questions, and it is now clear that knowing the answers to them could help us to improve mathematics education. A careful, detailed review of these studies is needed. One outcome of such a review will undoubtedly be specific guidance for further studies to refine and extend existing knowledge.

Locus of Control [20]

How much say should a student have in what he is to learn and how he is to learn it? Some educators urge that the locus of control over the curriculum should be shifted from the teacher (or, more generally, from the school system) to the learner.

Half of the studies to which I have references were easily available to me at the time of this writing. Of these, twice as many reported better student achievement in teacher controlled curricula than in the student situations.

Of course, this was not a randomly chosen sample of the studies, so the first order of business should be a careful review of all the investigations. Whether further research efforts should be called for will depend on the findings of this review.

This approach to teaching assumes that almost all students can achieve at a very high level. A general description of the underlying theory, as well as an annotated bibliography of related research up to 1970, was prepared by Block (17).

For the teaching of mathematics, two procedures have been tried. One is the mastery learning approach put forth originally by Bloom (18), and the other is the personalized system of instruction by Keller (19). The latter has been used almost exclusively at the college level.

Both versions seem to work quite well at the college level, although the economics of both seem to need further study. Very little study of these teaching procedures has yet been done at the secondary school level, and what results have been obtained are mixed. Studies at the elementary school level have been largely negative.

Further investigation of mastery learning at the college level seems needed, but it is doubtful that further study at the pre-college level would be useful at present.

Mathematics Laboratory [58]

A mathematics laboratory contains a variety of concrete objects--in particular, measuring devices--which can be used in mathematics instruction. The phrase "activity learning" is often used for an instructional program which makes much use of a mathematics laboratory.

Interest in mathematics laboratories became quite widespread in the latter part of the previous decade. Although they were hailed with a good deal of enthusiasm, the experimental studies of their effects on student achievement have not been very encouraging. While some studies have reported improved mathematics achievement resulting from mathematics laboratories, a substantially larger number of studies found no such advantage.

A review by Vance and Kieren (20) covers the research through 1970.

In the late fifties a number of educators asserted that instructional films and instructional TV would solve many of our educational problems by bringing the "master" teacher to millions of students instead of just one class of thirty students. Unfortunately, the validity of this very plausible idea turned out to be only moderate.

The review by Jamison, Suppes, and Wells (8) provides information on the general effectiveness of instructional TV and instructional radio as compared with conventional classroom instruction. Of the references I have collected to studies explicitly dealing with mathematics learning, about half are concerned with instructional TV. The remainder deal with, in decreasing order of frequency, instructional films, audio tapes, and overhead projectors.

The findings in all four areas are roughly the same: the alternative medium sometimes provides slightly better mathematics learning but just about as often provides no advantage. Only rarely is conventional classroom instruction significantly more effective than the alternative.

It is my belief that the reasons why these alternative media, except perhaps for the overhead projector, are not very widely used in this country are first that they are inconvenient to use and second that they involve some extra costs.

Of course, in countries where there is a shortage of qualified teachers, instructional TV and instructional radio can be, and are being, used effectively. But, until the delivery systems can be simplified and reduced in cost, further study of the effectiveness of these media is not a pressing need for us in the U.S.A.

Programmed Instruction [233]

Another innovative instructional procedure which came into prominence during the fifties was programmed instruction, sparked to no small extent by Skinner (21). Proponents of this innovation promised that it would enable all students to learn painlessly what only the ablest ones had learned before. While programmed instruction did not live up to this promise, some values which had not been anticipated by its proponents did eventually become apparent.

A general discussion of the values of programmed instruction is found in the Jamison, Suppes, and Wells (8) review.

About half the references I have found to studies specifically related to mathematics learning deal with technical questions about programs, e.g., the amount of new information to be included in each new step (large or small) or the nature of the student response to each step (constructed response or multiple-choice). Skinner's theory on which he based his recommendation for programmed instruction provided specific answers to many of these technical questions. Most of these answers have been shown to be wrong, and the very tight restraints placed by the original theory can be ignored. Nevertheless, there is a great deal of information in these studies which could be used by educators constructing new programs.

The remainder of the studies, for the most part, compare programmed instruction with conventional classroom instruction. The usual result is that students learn about as well from one procedure as from the other, and that neither is spectacularly better than the other.

The importance of this finding is that, unlike instructional TV and instructional radio, programs are easy and convenient to use. They require little or no teacher time, so they can be used for remedial or for enrichment purposes and as supplements to a standard program. Students seem to become bored on a steady diet of programs, so there is little possibility of their replacing teachers entirely, but programs clearly can be used in many ways to improve mathematics education.

What is needed now, in my opinion, is a careful review of all the studies which have been carried out so far. The results of such a review should be some quite explicit guidance to program construction on the one hand and to effective ways of using programs on the other. Any decision on further research on programmed instruction should be based on the results of this review.

It should be mentioned here that the research literature contains many references to studies which involved programmed instruction but which I did not include under this heading. These studies used programs as an instructional vehicle in studying other educational variables and were not intended to study programming per se.

A variety of ways of reinforcing or rewarding students for good performance in mathematics have been studied. About two-thirds of the studies investigated verbal reinforcement, for the most part praise. A frequently studied variable was the frequency with which praise was provided.

Of those reports which I have looked at in some detail, the results were mixed. There were enough positive results, however, to indicate that further studies, after a careful review of what has already been done, would be worth the effort.

The remainder of the studies dealt with material rewards, ranging from candy and small toys to extra time at recess. The results, especially for disadvantaged young children, have been sufficiently encouraging to suggest strongly that further investigation be carried out.

Team Teaching [20]

In team teaching, two or three teachers pool one class each and team together to teach the whole set of students. The idea behind this procedure is that each teacher will utilize his own particular teaching strengths so that the total effect on the students will be better than that of any one individual teacher.

Armstrong (22) has provided a general review of the effects of team teaching on student achievement. Together with some of my students, I (23) reviewed those studies specifically including measurement of mathematics achievement. This latter review makes it clear that team teaching provides no advantage over conventional teaching. In fact, the findings are so clear and unequivocal that any further investigation of this teaching procedure would seem to be a waste of effort.

Tests [20]

Although classroom teachers use tests primarily for diagnostic or for evaluation purposes, there is now enough empirical information to suggest that increased frequency of testing leads to improved student performance. The evidence also suggests, but as yet nowhere near conclusively, that easy tests are more effective for this purpose than difficult ones and that increasing the frequency of testing does not lead to increased test anxiety.

This is clearly a topic on which more detailed information is needed.

Time [36]

Under this heading I have placed those studies which have investigated various aspects of the allocation of classroom time. Some studies were concerned with different ways of allocating a fixed amount of time. Thus, for example, four 50-minute periods could be compared, with respect to student achievement, with two 100-minute periods.

Other studies investigated the effect on student achievement of an increase in the amount of classroom time devoted to mathematics. Not surprisingly, more time usually results in greater achievement. (Curiously enough, however, one study reported greater achievement in computation in a 40-minute class session than in a 50-minute session.)

About a third of the studies were concerned with a model of school learning put forward by Carroll (24) in which the time required to learn a particular topic may vary from student to student.

The kinds of information provided by these kinds of studies could be quite useful to mathematics educators. The studies already completed should be reviewed and the information in them codified. If further studies are needed, that fact should become apparent after such a review.

Carroll's model is an intriguing one. Further attempts to fit this model to actual mathematics learning situations should be encouraged.

Tutoring [51]

Concern for the education of the disadvantaged was a feature of the sixties. One of the common procedures used in attempting to help low achievers, who were of course most numerous among the disadvantaged, was tutoring. A large number of studies were aimed at investigating the benefits of tutoring in general and of particular tutoring procedures in particular.

As a result of this work, we now know that tutoring is often effective. What is needed now is a careful review of all these studies to look for answers to such questions as: "Which tutoring procedures are most effective for particular sets of students?" ". . . for particular mathematical objects?" (Tutoring for computational skills and tutoring for understanding of mathematical concepts or principles may require different techniques.) "Are there training procedures which increase tutor effectiveness?"

Further investigation in this general area should probably await the completion of such a review.

Summary Comments

To educators looking for ways to improve mathematics education, the variables reviewed in this chapter are especially important. Like curriculum variables, and unlike teacher, student, and most environment variables, they can easily be manipulated. When one of these variables is demonstrated to have a beneficial effect on mathematics learning, it can be put into use by the classroom teacher. When one of these variables is shown not to have an effect on mathematics learning, it can be tuned out by the classroom teacher.

It is encouraging, therefore, to see that a good deal of information already exists about many of the variables and only awaits careful review and codification. Of course some of these variables, such as team teaching, will turn out not to be helpful. But others, such as tests, on the basis of preliminary partial reviews at least hold out promise of turning out to be quite helpful.

In short, it is clear that it would be very worthwhile to collect and organize the vast amount of information about instructional variables which is now buried in scattered dissertations, research reports, and journal articles.

Bibliography

1. Calculator Information Center. Research on Hand-held Calculators. Reference Bulletin No. 2, ERIC, Ohio State University, 1977.

2. Calculator Information Center. Research on Desk Calculators. Reference Bulletin No. 3, ERIC, Ohio State University, 1977.

3. Dunkin, M. J., and Biddle, B. J. *The Study of Teaching.* New York: Holt, Rinehart, and Winston, 1974.

4. Rosenshine, B. Teaching Behaviors and Student Achievement. London: National Foundation for Educational Research in England and Wales, 1971.

5. Rosenshine, B., and Furst, N. Research on Teacher Performance Criteria. In Smith, B. Othanel (Ed.), *Research in Teacher Education.* Englewood Cliffs, NJ: Prentice Hall, 1971.

6. Heath, R. W., and Nielson, M. A. The Research Basis for Performance-Based Teacher Education. *Review of Educational Research,* Vol. 44 (1974) pp. 463–484.

7. Shavelson, R. and Dempsey-Atwood, N. Generalizability of Measures of Teacher Behavior. *Review of Educational Research,* Vol. 46 (1976) pp. 553–611.

8. Jamison, D., Suppes, P., and Wells, S. The Effectiveness of Alternative Instructional Media: A Survey. *Review of Educational Research,* Vol. 44 (1974) pp. 1–67.

9. Suydam, M. N. The Use of Computers in Mathematics Education: Bibliography. ED 077 733.

10. Canadian Council for Research in Education. Canadian Experience with the Cuisenaire Method. 1964.

11. Beberman, M. *An Emerging Program of Secondary School Mathematics.* Cambridge, MA: Harvard University Press, 1958.

12. Ausubel, D. P. Learning by Discovery—Rationale and Mystique. *Bulletin of the National Society of Secondary School Principals,* Vol. 45 (1961) pp. 18–58.

13. Shulman, L. S., and Keislar, E. R. (Eds.). *Learning by Discovery: A Critical Appraisal.* Chicago: Rand McNally, 1966.

14. Miller, R. L. Individualized Instruction in Mathematics: A Review of the Research. *Mathematics Teacher*, Vol. 69 (1976) pp. 345-351.

15. Schoen, H. L. Self-paced Mathematics Instruction: How Effective Has It Been? *Arithmetic Teacher*, Vol 23 (1976) pp. 90-96.

16. Schoen, H. L. Self-paced Mathematics Instruction: How Effective Has It Been in Secondary and Post-Secondary Schools? *Mathematics Teacher*, Vol. 69 (1976) pp. 352-357.

17. Block, J. H. (Ed.). Mastery Learning: Theory and Practice. New York: Holt, Rinehart, and Winston, 1971.

18. Bloom, B. S. Mastery Learning. In 17, above.

19. Keller, F. S. "Goodbye Teacher. . . ." *Journal of Applied Behavior Analysis*, Vol. 1 (1968) pp. 78-89.

20. Vance, J. H., and Kieren, T. E. Laboratory Settings in Mathematics: What Does Research Say to the Teacher? *Arithmetic Teacher*, Vol. 18, (1971) pp. 585-589.

21. Skinner, B. F. The Science of Learning and the Art of Teaching. *Harvard Educational Review*, Vol. 24 (1954) pp. 86-97.

22. Armstrong, D. G. Team Teaching and Academic Achievement. *Review of Educational Research*, Vol. 47 (1977) pp. 65-86.

23. Begle, E. G., et al. Review of the Literature on Team Teaching in Mathematics. Teacher Corps Mathematics Work/Study Team, Working Paper No. 3. ED 138 548.

24. Carroll, J. B. A Model of School Learning. *Teachers College Record*, Vol. 64 (1963) pp. 723-733.

Illustrative Research Reports

Austin, J. D. An Experimental Study of the Effects of Three Instructional Methods in Basic Probability and Statistics. *Journal for Research in Mathematics Education,* Vol. 5 (1974) pp. 146-154.

Bartz, W. H., and Darby, C. L. The Effects of a Programmed Textbook on Achievement Under Three Techniques of Instruction. *Journal of Experimental Education,* Vol. 34 (1966) pp. 46-49.

Begle, E. G. Time Devoted to Instruction and Student Achievement. *Educational Studies in Mathematics,* Vol 4 (1971) pp. 220-224.

Crosby, G., and Fremont, H. Individualized Algebra. *Mathematics Teacher,* Vol. 53 (1960) pp. 109-112.

Denny, T., Paterson, J., and Feldhusen, J. Anxiety and Achievement as Functions of Daily Testing. *Journal of Educational Measurement,* Vol. 1 (1964) pp. 143-147.

Dysinger, D. W., and Bridgman, C. S. Performance of Correspondence-Study Students. *Journal of Higher Education,* Vol. 28 (1957) pp. 387-388.

Fisher, M., et al. Effects of Student Control and Choice of Engagement in a CAI Arithmetic Task in a Low-Income School. *Journal of Educational Psychology,* Vol. 67 (1975) pp. 776-783.

Gray, R. F., and Allison, D. E. An Experimental Study of the Relationship of Homework to Pupil Success in Computation with Fractions. *School Science and Mathematics,* Vol. 71 (1971) pp. 339-346.

Griggs, S. A. An Experimental Program for Corrective Mathematics in Schools for Socially Malajusted (sic) and Emotionally Disturbed. *School Science and Mathematics,* Vol. 76 (1976) pp. 377-380.

Hanna, G. S. Effects of Total and Partial Feedback in Multiple-Choice Testing upon Learning. *Journal of Educational Research,* Vol. 69 (1976) pp. 202-205.

Hatfield, L. L., and Kieren, T. E. Computer-Assisted Problem Solving in School Mathematics. *Journal for Research in Mathematics Education,* Vol. 3 (1972) pp. 99-112.

Heitzman, A. J. Effects of a Token Reinforcement System on the Reading and Arithmetic Skills Learnings of Migrant Primary School Pupils. *Journal of Educational Research,* Vol. 63 (1970) pp. 455-458.

Helwig, C., and Griffin, S. Ninth Graders Teach Arithmetic to Fifth
    Graders: An Experiment. *School Science and Mathematics,* Vol. 74
    (1974) pp. 13-15.

Houston, W. R., and DeVault, M. V. Mathematics In-Service Education:
    Teacher Growth Increases Pupil Growth. *Arithmetic Teacher,*
    Vol. 10 (1963) pp. 243-247.

Jacobs, J. N., and Bollenbacher, J. Teaching Seventh-Grade Mathematics
    by Television. *Mathematics Teacher,* Vol. 53 (1960) pp. 543-547.

Jones, T. The Effect of Modified Programmed Lectures and Mathematical
    Games upon Achievement and Attitude of Ninth-Grade Low Achievers
    in Mathematics. *Mathematics Teacher,* Vol 61 (1968) pp. 603-607.

Lackner, L. M. Teaching of Limit and Derivative Concepts in Beginning
    Calculus by Combinations of Inductive and Deductive Methods.
    Journal of Experimental Education, Vol. 40 No. 3 (1972) pp. 51-56.

Maltbie, A., Savage, R. G., and Wasik, J. L. The Operation and Evalua-
    tion of a Proctorial System of Instruction in Mathematics.
    American Mathematical Monthly, Vol. 81 (1974) pp. 71-78.

O'Loughlin, T. Using Electronic Programmable Calculators (Mini-
    Computers) in Calculus Instruction. *American Mathematical Monthly,*
    Vol. 83 (1976) pp. 281-283.

Paige, D. D. A Comparison of Team Versus Traditional Teaching of
    Junior High School Mathematics. *School Science and Mathematics,*
    Vol. 67 (1967) pp. 365-367.

Schultz, E. W. The Influence of Teacher Behavior and Dyad Compatibility
    on Clinical Gains in Arithmetic Tutoring. *Journal for Research
    in Mathematics Education,* Vol. 3 (1972) pp. 33-41.

Suppes, P., Jerman, M., and Groen, G. Arithmetic Drills and Review
    on a Computer-Based Teletype. *Arithmetic Teacher,* Vol. 13 (1966)
    pp. 303-309.

8

Tests

Mathematics tests are used in almost every aspect of mathematics education. Teachers use them to evaluate the progress of their students. Tests are used in evaluations of local, state, and national mathematics education programs, and also in evaluation of new curriculum materials, such as textbooks, workbooks, and mathematics films.

Almost any piece of research in mathematics education involves the use of tests, since the object of such research is to see if something has an effect on students' ability to do mathematics, which can be determined only by administering tests to the students.

Classroom teachers usually use tests prepared by themselves or, sometimes, in collaboration with their colleagues. For other purposes, "standardized" tests are often used. These are tests developed and sold by professional organizations. For these tests, "norms" are available, i.e., information about the scores obtained on the test by a representative set of students. Braswell (1) has prepared a list of the standardized mathematics tests currently available in this country.

There are two important sources of information about the quality of standardized tests. The first is a series of reviews compiled by Buros (2). The second is a series of reviews prepared by the Center for the Study of Evaluation (3).

For research purposes, standardized tests are sometimes not appropriate. Thus, for example, the goals of a particular educational process being studied may not be exactly matched by any standardized test. Consequently, a large number of special purpose tests have been developed during the past two decades.

135

This chapter is restricted to studies which concentrate on the development and characteristics of both standardized and special purpose tests.

Various kinds of information are provided in these studies. In some cases, only statistical information such as item difficulties, reliability, or validity are discussed. In some other cases, comparisons with other similar tests, usually in the form of correlations, are provided. Studies which include information about tests but which are more concerned with the information obtained, through use of the tests, about other educational variables are taken up in other chapters.

Affective Tests [16]

Student attitudes are often of interest to mathematics educators. Among the attitudes measured by the tests in these studies are those toward mathematics in general as well as toward various areas of mathematics (e.g., arithmetic, geometry). Also, a number of measures of mathematical self-concept exist.

In addition to the sixteen research reports on attitude tests which I have located, NLSMA Reports Nos. 1, 2, and 3 reproduce a number of affective tests which were used in NLSMA, and Reports Nos. 4, 5, and 6 provide statistical information about these tests. In addition, NLSMA Report No. 9 deals, among other things, with some attitude tests administered to NLSMA teachers.

If student attitudes are as important as most educators seem to believe, then a great deal more needs to be done in developing affective tests. For example, there are many different aspects of arithmetic. But, while we have tests of students' attitudes toward arithmetic in general, we have no tests which tell us which students prefer fractions to whole numbers, or word problems to rote computation, etc.

Studies of such more specialized attitudes will require specialized attitude tests, and if such tests were developed and made available, then research on such attitudes would go forward more quickly.

Most standardized tests are "norm-referenced," which is to say that they are meant to be used to identify an individual's performance in relation to the performance of others on the same test. Some recent developments in education, in particular "mastery learning," discussed in Chapter Seven, require a different kind of test, namely ones which are intended to measure an individual's status with respect to an established standard of performance. Such tests are called "criterion-referenced."

Two recent, general bibliographies on criterion-referenced testing have been prepared by Ellsworth and Franz (4) and by Porter (5).

There seems to have been more discussion of criterion-referenced tests than there has been development of them. However, whether more such tests need to be developed in the future will depend on the results of studies, such as those in mastery learning, in which they are used.

Item Sampling [16]

This term refers to a particular procedure for administering tests. The phrase "matrix sampling" is also used. The procedure consists in administering each item of a test not to the whole student population being measured but rather to a sample of the students. The point of the procedure is that if the samples are chosen randomly and if they are large enough, then the average score on the item for the sample will be a close approximation to the average score which the entire population would have obtained. This means that if the student population being tested is large enough, then each student need appear only in a few item-samples and hence the testing time for each student is much reduced. Alternatively, without increasing the normal testing time, a longer test (or larger test battery) can be used. In either case, the statistical information provided by this procedure will be close enough for all practical purposes to that which would have been obtained by administering the entire test to the entire population.

There are many cases where the student population is large enough to allow this procedure to be used. In particular, item sampling is usually used in the evaluation of educational programs in large school systems and at the state or national level.

All of the studies in this area were concerned with the technical aspects of the procedure. There is no indication that the procedure requires special kinds of tests.

Mathematics Tests [81]

About half of these studies were concerned with standardized tests and computed correlations between selected pairs of them or investigated subsets of the sets of test items or examined their validities in terms of correlations with student performance in mathematics courses.

The remainder were concerned with special purpose tests, usually constructed by the experimenter. In general, these studies did little more than compute the usual statistics for these tests.

Probably more work in both these directions would be useful, but until the existing reports have been carefully reviewed and their findings compared, it will not be possible to be specific about future research.

Test Administration [38]

In administering a mathematics test, a number of different variables can be manipulated. The effects of several of them on student scores have been studied. Examples of such variables are: answer format (multiple-choice vs. constructed response); answer sheet format (in test booklet vs. separate answer sheet); coaching (previous exposure to item format vs. no previous exposure); time limit (vs. no time limit); previous notice of test (vs. no notice); use of book (open vs. closed book test); race or sex of examiner; and item sequence (easy to hard vs. random sequence).

In many cases, the variable being studied did have an effect on student scores, although these effects did not always result in changes in the rank ordering of the students. Since test results can have strong effects on the well-being of students, it is clear that the available information should be carefully reviewed and that further research should be carried out to clarify the effects of all these variables.

Of course, there are some who argue that tests results, since tests are fallible, should never be used in making decisions about individual students. This argument does not seem to hold up. In general, some information is better than none. We should devote our efforts to reducing the fallibility of tests rather than eliminating them.

Test and Item Characteristics [29]

The boundary between this and the previous sub-topic is somewhat ambiguous since changes in test or item characteristics can be made before the administration of the test.

Some of the variables studied in these reports were: item wording; scoring methods (should account be taken of guessing?); item format (include extraneous information?); number of choices for multiple-choice items; ratio of high cognitive level to low cognitive level items.

The comments made about test administration apply here without change.

Test Construction [8]

While numerous articles have been written telling how to construct a mathematics test, I have located only a few which provided numerical information about the development of a specific test.

Lippey (6) has compiled a bibliography on the use of computers in test construction. Romberg and Wilson provide an account of the development of the NLSMA tests in NLSMA Report No. 7.

Those who agree with me that tests are an indispensable tool in mathematics education will also agree with me that we need to know more about the test construction process.

Summary Comments

There exist numerous texts and journal articles, at various levels of sophistication, on test theory, test construction, test administration, and the uses of tests in general. Anyone interested in the role of mathematics tests in mathematics education should be aware of this literature and acquainted with at least some of it.

A detailed inspection of each of the empirical reports surveyed in this chapter would without doubt provide us with a wealth of important information about mathematics tests. More importantly, only such a review can provide the guidance we need for further research.

Bibliography

1. Braswell, J. S. Mathematics Tests Available in the United States. National Council of Teachers of Mathematics. ED 121 616.

2. Buros, O. *Seventh Mental Measurements Yearbook*. Highland Park, NJ: The Gryphon Press, 1972.

3. Center for the Study of Evaluation. CSE Secondary Schools Test Evaluation. University of California at Los Angeles.

4. Ellsworth, R. A., and Franz, C. Bibliography on Criterion Referenced Measurement. ED 115 699.

5. Porter, D. E. Criterion Referenced Testing: A Bibliography. ED 117 195.

6. Lippey, G. Bibliography on Computer-Assisted Test Construction. ED 095 909.

Illustrative Research Reports

Beck, M. D. Achievement Test Reliability as a Function of Pupil-
Response Procedures. *Journal of Educational Measurement,*
Vol. 11 (1974) pp. 109-114.

Cahen, L. S., Romberg, T. A., and Zwirner, W. The Estimation of
Mean Achievement Scores for Schools by the Item-Sampling Technique.
Educational and Psychological Measurement, Vol. 30 (1970) pp. 41-60.

Coppedge, F. L., and Hanna, G. S. Comparison of Teacher-Written and
Empirically Derived Distractors to Multiple-Choice Test Questions.
Journal for Research in Mathematics Education, Vol. 2 (1971)
pp. 299-303.

Hively, W., II, Patterson, H. L., and Page, S. H. A "Universe-Defined"
System of Arithmetic Achievement Tests. *Journal of Educational
Measurement,* Vol. 5 (1968) pp. 275-290.

McCallon, E. L., and Brown, J. D. A Semantic Differential Instrument
for Measuring Attitude Toward Mathematics. *Journal of Experimental
Education,* Vol. 39 No. 4 (1971) pp. 69-72.

Silver, J., and Waits, B. Multiple-Choice Examinations in Mathematics,
Not Valid for Everyone. *American Mathematical Monthly,* Vol. 80
(1973) pp. 937-942.

Weaver, J. F. Disparity in Scores from Standardized Arithmetic Tests.
Arithmetic Teacher, Vol. 9 (1962) pp. 96-97.

9

Problem Solving

Mathematics is a lot of fun for a small number of individuals. For even a smaller number mathematics provides a profound aesthetic experience.

If that were the whole story, it would not be possible to justify the emphasis given to mathematics in our school programs. The real justification for teaching mathematics is that it is a useful subject and, in particular, that it helps in solving many kinds of problems. In this chapter we survey the research on mathematical problem solving.

A general review of problem-solving research, not restricted to mathematics, was prepared by Davis (1). A general review of the literature on mathematics problem solving has been provided by Kilpatrick (2). His review was prepared in 1969, but the general shape of the area seems not to have changed much since then. A more recent discussion by Simon (3) is also of value. The use of computers in problem solving has been reviewed by Hatfield (4).

Ability [63]

Studies of problem-solving ability are similar to those of general mathematical ability, which were surveyed in Chapter Five, the difference being that here the criterion tests concentrate on problem solving rather than mathematical knowledge in general. The same research techniques were used. In many cases, correlations were sought between specific cognitive abilities and proficiency in solving problems. In some others, contrasts between the patterns of cognitive abilities of good and poor problem solvers were studied.

A review of this research area has been prepared by Trimmer (5).

These efforts to characterize problem-solving ability in terms of more specific cognitive abilities have been no more successful than the attempts to characterize mathematical ability. In fact, some of the more recent studies suggest that problem-solving ability is probably not a unitary trait and that different mixtures of abilities are needed for different classes of problems.

Looked at in retrospect, this last suggestion, if true, should not surprise us. It points out that simplistic efforts to improve our students' problem-solving abilities will not be enough. The task is much more complex than that.

Instructional Programs [36]

Despite our lack of any demonstrated theoretical understanding of problem-solving ability, a number of instructional programs have been devised which were intended to increase students' ability to solve mathematical problems. It is interesting that the evaluations of these programs sometimes had statistically significant results. Further investigations along these lines might be reasonable, preceded of course by a careful analysis of what has already been done.

Problem Format [25]

The format of a problem can be changed in various ways. For example, a diagram may or may not accompany the problem. The vocabulary and grammatical complexity can be modified. The symbol-to-word ratio can be increased or decreased. The question can be placed at the beginning or end of the problem.

The effects of such format variables on problem difficulty have been studied to some extent. There is no doubt about there being a significant effect in some cases, but not enough has yet been done to allow drawing any broad conclusions.

If certain format changes turn out to reduce problem difficulty, then it would be worth experimenting with attempts to teach students to make the format changes themselves. Consequently, a good deal more work in this area is needed.

Problem Structure [26]

Mathematical problems differ among themselves with respect to their structures. Both the kinds of operations and the minimum number of operations needed to solve the problem can vary from one problem to another. We would probably say that an "original" in geometry and a problem asking about the roots of a quadratic equation have different structures.

The effects of problem structure on problem difficulty has received some study. As in the case of problem format, no broad generalizations came out of these findings. It is hard to say, however, whether further work in this area would be worthwhile. After all, there is nothing the problem solver can do to alter the structures of the problems posed him.

Strategies [75]

A substantial amount of effort has gone into attempts to find out what strategies students use in attempting to solve mathematical problems. The problems investigated have ranged from simple arithmetic ones suitable for primary students to complex algebraic and geometric problems meant for senior high school students.

Some of the most interesting of these studies have been computer simulations of putative problem-solving strategies. Further efforts along these lines would be welcome.

No clear-cut directions for mathematics education are provided by the findings of these studies. In fact, there are enough indications that problem-solving strategies are both problem- and student-specific often enough to suggest that hopes of finding one (or few) strategies which should be taught to all (or most) students are far too simplistic.

At least three different kinds of questions have been asked which can be placed under this heading. "Does it help students if they verbalize during the problem-solving process?" "Is there a high correlation between reading ability and problem-solving ability?" "Does a study of mathematics vocabulary improve problem solving?"

The findings so far do not provide clear-cut answers for any of these questions, but the indications, especially for the last question, are sufficiently favorable that rather extensive further research is indicated.

Summary Comments

This brief review of what we know about mathematical problem solving is rather discouraging. Compared to the importance of the topic, the amount of factual information that is available to us is quite small. Even more discouraging is that little interesting research is going on at the present. In particular, interest in the information-processing approach to problem solving embodied in the computer simulation studies of the early sixties (see (6) for an illustration) seems to have largely dissipated.

Bibliography

1. Davis, G. The Current Status of Research and Theory in Human Problem Solving. ED 010 506.

2. Kilpatrick, J. Problem Solving and Creative Behavior in Mathematics In J. W. Wilson and L. R. Carry (Eds.). *Reviews of Recent Research in Mathematics Education.* Studies in Mathematics, Vol. 19 Stanford, Calif.: School Mathematics Study Group, 1969. (Note: A shortened version of this review entitled Problem Solving in Mathematics, appeared in *Review of Educational Research,* Vol. 39 (1969) pp. 523–534.

3. Simon, H. A. Learning with Understanding. ED 113 206.

4. Hatfield, L. L. Computer-Extended Problem Solving and Enquiry, II. ED 077 732.

5. Trimmer, R. G. A Review of the Research Relating Problem Solving and Mathematics Achievement to Psychological Variables and Relating These Variables to Methods Involving or Compatible with Self-Correcting Manipulative Mathematics Materials. ED 092 402.

6. Newell, A., and Simon, H. A. *Human Problem Solving.* Englewood Cliffs, N. J.: Prentice-Hall, 1972.

Arter, J. A., and Clinton, L. Time and Error Consequences of Irrelevant Data and Question Placement in Arithmetic Word Problems II: Fourth Graders. *Journal of Educational Research,* Vol. 68 (1974) pp. 28–31.

Chase, C. I. The Position of Certain Variables in the Prediction of Problem-Solving in Arithmetic. *Journal of Educational Research,* Vol. 54 (1960) pp. 9–14.

Dahmus, M. E. How to Teach Verbal Problems. *School Science and Mathematics,* Vol. 70 (1970) pp. 121–138.

Jerman, M. Some Strategies for Solving Simple Multiplication Combinations. *Journal for Research in Mathematics Education,* Vol. 1 (1970) pp. 95–128.

Johnson, H. C. The Effect of Instruction in Mathematical Vocabulary Upon Problem Solving in Arithmetic. *Journal of Educational Research,* Vol. 38 (1944) pp. 97–110.

10

Reflections and Conclusions

Critical Variables for Review

In the body of this report I have suggested for each of a large
number of variables that a comprehensive review of all the existing
relevant empirical literature would be worthwhile. The number of suggestions is quite large, so I would like to give here some indication
of my own priorities. I have therefore chosen eight topics to which I
assign top priority for comprehensive reviews of the literature. I am
not willing to specify any degrees of relative importance among these
critical variables, so I present them in the order in which they
appeared in the preceding chapters.

Goals for Mathematics Education

As was pointed out in Chapter Two, there is, on the one hand, no
unique set of goals that we have agreed on in this country and, on the
other hand, increasing pressure for the specification of goals. Part
of this pressure comes from the demands from the general public for
accountability on the part of school systems and teachers. Another
comes from the increasing tendency to mandate a specified level of competence in mathematics (and of course in other subjects) as a prerequisite for graduation from high school. Perhaps less visible to most,
there is pressure from educational researchers for clearer statements
of goals. It is now quite well realized that in a well-designed
research project, the choice of criterion variables should not be
decided merely on the basis of the availability of a particular standardized test.

While there is a great deal of diversity in the sets of goals which have been chosen by different sets of individuals, there is at the same time, I believe, a good deal in common among these sets of goals. I believe that it would be very helpful to teachers, school administrators, parents, and others if a brief brochure could be prepared which would make both these facts clear.

There are many different formats which such a brochure might use, and the one which I sketch is merely an illustration. I suggest that half a dozen or so sets of goals be presented, each of which has been adopted by a responsible group, such as a substantial school system, a state department of education, or a national survey. These sets of goals, placed side by side, should make it easy to illustrate the two points made above: diversity and at the same time similarity.

A school system, for example, attempting to specify its goals for mathematics education might, as a result of the discussion of the set of goals presented in the brochure, have its attention called to a goal which it might otherwise have inadvertently overlooked. Conversely, it might be led to question one of its pet goals if it turned out that no one else valued it. And most important, it would emphasize that those setting goals need not be slavish imitators of any one else.

While those preparing such a brochure will probably want to review most of the items included in the Bibliography of Chapter Two, probably only a selected small part of the results need be incorporated in the final document.

Meaningful Instruction

While this phrase is still somewhat ambiguous, the great success of "meaningful" curriculum materials argues very strongly for a careful, extensive review of the literature. I would hope that one result of such a review would be a more operational definition of the term which would provide better directions for the construction of meaningful instructional materials in the future.

Class Size

There should be enough in the literature by now to settle the question of the relationship between mathematics class size and student

learning. But most discussions of this topic seem to ignore the literature. The facts should be brought out.

Programmed Instruction

As was pointed out, programmed materials are easy to use, in distinction to films and TV. While it is unlikely that the bulk of mathematics instruction will ever be based on programmed units, they can
nevertheless be used in so many supplementary ways in the classroom
that it would be wise to gather together everything that is known about
the effects of such materials.

Tests

Although the literature is not very extensive, the possibility that
frequent easy tests might increase student learning is so intriguing
that this part of the literature deserves a careful review. This is an
instructional device which can be easily and widely implemented if the
findings of the review turn out to be favorable.

Tutoring

This seems to be one of the most effective procedures for improving
mathematics instruction for disadvantaged or low achieving students.
The literature is quite extensive and it would seem wise to cull from
it whatever we can about optimal methods of tutoring, as well as optimal
methods of training tutors, for some sort of training does seem to be
necessary.

Test Administration and Test Characteristics

We noted that variations in administration procedures or in test
formats can lead to differences in student test scores. Since these
scores play a significant part in decisions both about the students
themselves and about curricular methods and materials, it is very important that everything we know about these two aspects of testing be made
available both to classroom teachers and to school administrators,
educational researchers, etc.

Problem-Solving Strategies

While the ability to solve mathematical problems is an essential goal of mathematics education, the literature on this general topic is not very encouraging. However, there have been enough studies of the strategies which students use in attacking mathematical problems so that a careful review of this part of the literature might provide some useful suggestions and in any case should indicate the most profitable directions for further research.

I do not wish to suggest that a careful review of the empirical literature about any of the above critical variables will necessarily provide us with a firm indication of the value, or lack of value, of that topic to mathematics education. I suspect that it will in some cases, while in other cases it may merely lead to a clearer understanding of the proper direction for further research.

In any case, it should be noted that each of these variables could be manipulated fairly easily by the classroom teacher, the school administrator, or the textbook author. Anything positive which results from reviews of these topics could be put to use, widely and quickly, to improve mathematics education.

Critical Variables for Research

In addition to suggesting, in the preceding chapters, that comprehensive literature reviews would be appropriate for many educational variables, a high priority subset of which is listed above, I also identified on the basis of brief surveys of literature a large number of topics for which further research would probably be appropriate. I have chosen five of these as the ones that are most critical and to which I would give highest priority. It goes without saying, however, that further research should be preceded by a careful review of the available literature. It would be unfortunate to overlook any promising leads which such an overview might turn up, and it would be even more unfortunate to repeat mistakes which previous research has uncovered.

The Relationship Between Teacher Knowledge of Subject Matter and Student Achievement

As I pointed out in Chapter Three, it is not necessarily the case that the more the teacher knows the more the students learn. But until we have a better understanding of the relationship between teacher knowledge and student learning, we do not know how much subject matter instruction needs to be built into our teacher training programs. We have no rational defense for the number of mathematics courses we ask prospective elementary school teachers to take and no rational grounds for asking for more. And the situation with respect to training programs for secondary teachers is equally uncertain. We do not know what the best topics are for in-service training programs.

To obtain the needed information for a broad range of mathematical topics and a wide spectrum of student interests and abilities will require a very large and well-coordinated research effort. But the value of the outcome should be more than enough to justify the cost and effort.

Drill

There seems to be no doubt that a certain amount of drill on newly learned mathematical topics improves student mathematics achievement. But many questions remain about the timing (after how much meaningful instruction?), the amount, the spacing (massed or distributed?), etc. of drill. Nor do we know how the answers to these questions vary with mathematical topic, cognitive level, and student ability and interest.

Answers to these questions could lead to substantial increases in the efficiency of our mathematics educational programs. It would take another large, coordinated research program to obtain these efforts, but again the benefits would easily justify the cost.

Expository Teaching of Mathematical Objects

I pointed out in Chapter Seven that the experiments comparing expository with discovery teaching had not demonstrated a strong advantage for either one. On the other hand, there is no unambiguous definition of "discovery," and as a result it is difficult, if not impossible, to design experiments involving discovery teaching which have any hope of being replicable or of yielding generalizable results. For this reason I limit this recommendation to expository teaching.

There has been some, but not nearly enough, study of the expository teaching of mathematical concepts. Variables which have been investigated are: the number of examples and non-examples provided, the number of problems for the learner to work requiring the use of the concept, and the degree of emphasis on subordinate concepts. Detailed information on the importance of these variables, for different subject matters within mathematics and for different student ability levels, is needed. Very little study has been devoted to the expository teaching of mathematical operations or mathematical principles. A substantial research effort is essential.

Acceleration

We can be fairly certain that our brightest students will profit, and no one else will be harmed, if they are separated from the rest of the student body and given a special mathematics program. However, we are not yet sure what kind of a program to provide--acceleration or enrichment. My guess is that for the very top layer of students acceleration is preferable but that, perhaps, for the next layer down either procedure is satisfactory. In any case, since our very brightest students do constitute a priceless national asset, this question needs to be answered.

Predictive Tests

There will always be some choice points in the mathematics program of any student. The most prominent of these comes sometime during the junior high school years when a decision has to be made whether the student should or should not attempt a formal course in algebra. Since an erroneous decision, in either direction, is damaging to the student, we need to improve the quality of the information we use in arriving at these decisions. One commonly used piece of information comes from tests which "predict" performance in the algebra course. There are numerous such tests available, but none of them are as discriminating as

we would like. Research aimed at improving these predictive tests could
have valuable results for a very large number of our students.

In the same way, we need more accurate tests to be used in deciding
which students should be separated from the bulk of the student body and
given either enrichment materials or acceleration.

There are probably other choice points where improved predictive
tests would be helpful, but the two areas mentioned above seem to me to
be by far the most important and the most deserving of extensive research.

General Comments

Now I wish to turn to some general comments.

First, as the preceding chapters have so amply demonstrated the
amount of factual information about mathematics education is extremely
large. There are many thousands of doctoral dissertations, journal
articles, and ERIC documents which contain such information. But, unfortunately, I must again remind the reader that this information is
not easily available. It is widely scattered. Before we can make use
of the information, it must be pulled together, organized, and made
available in convenient form. The kind of review of the literature
which I carried out for ability grouping (Chapter Six (1)) is one way to
make information available. There may well be others. In any case, it
seems foolish to ignore this treasury of information.

Of course, not all these packets of information are equally valuable. Some may turn out to be worthless. We do know that many of them
are valuable, and we will never know about the others until we dig them
out, inspect them carefully, and compare them with other related lists
of information.

In short, a substantial effort to review as much of the empirical
literature as possible is badly needed and as soon as possible. The
eight high priority topics listed above are just those that combine
a high probability of positive findings with high educational payoff.
They come nowhere near exhausting the raw material available to us.

A second comment looks more toward the future than toward the past.
One thing that struck me very forcefully in the course of reading or
even skimming the literature surveyed in the preceding chapters is that

the research efforts and empirical studies which have been carried out were, for the most part, uncoordinated. Very seldom have I found two studies of the same variable which used either the same measuring instruments or the same kinds of students. Almost never have I observed experiments being replicated.

As a result of this lack of coordination, it is harder, and requires many more bits of information, to reach firm conclusions about the effects of particular educational variables than would have been the case if more coordination had been practiced. Nothing can be done about the past, but I would hope that future research would be planned to include more coordination and thus an increased efficiency for the whole research endeavor.

My final general comment is that this survey of the empirical literature left me feeling quite depressed. There are two reasons for this feeling. On the one hand, I do not see that we have any substantial body of knowledge about mathematics education that we can build on. Thus, for example, I attended all three of the International Congresses on Mathematics Education, and the last two disappointed me greatly. At Exeter, no one talked along the lines of: "Here is something about mathematics education that we know now but did not know at the time of the Lyon Congress." And at Karlsruhe, no one pointed out knowledge that had been gained since the Exeter Congress. Of course, some opinions changed from Congress to Congress, and fresh individuals were called on to express these opinions, but new knowledge did not appear.

On the other hand, this lack of a solid knowledge base is in part explained by and at the same time helps to explain the fact that our research efforts have not been based on any broad theoretical foundations. It is true that each author of a research report tries to present some theoretical rationale for his research, but these rationale rarely go beyond references to previous studies or to narrow hypotheses which someone has proposed. Even when studies of a single narrow topic are reviewed, the rationales for the various studies rarely, if ever, fit together as parts of a single theoretical structure.

But, while I am depressed, I am not despondent. I do believe that there exists, in the literature, a solid body of information about mathematics education. And I do believe that once this information is dug out and organized, then it will begin to suggest testable theories and at the same time will provide a template against which tentative theories can be tested.

But unless this information is extracted from the scattered literature, organized, and made widely available, I predict that future International Congresses will be much like the past ones: new faces, new opinions, but very little new knowledge.

Biographical Sketch

Edward Griffith Begle was born 27 November 1914 in Saginaw, Michigan. He attended the University of Michigan, where he studied under Raymond L. Wilder and developed an interest in topology. He received his A.B. in mathematics from Michigan in 1936 and his M.A. in 1937, earning membership in Phi Beta Kappa and Sigma Xi. Then he went to Princeton University, where he obtained his Ph.D. in 1940 under the direction of Solomon Lefschetz. His dissertation topic was locally connected spaces and generalized manifolds.

Begle taught mathematics at Princeton during the 1940–41 academic year. The following year he returned to Ann Arbor as a National Research Council Fellow, and then in 1942 he went to Yale University, where he remained until 1961, rising through the ranks of Instructor to Associate Professor in the Department of Mathematics. While on the Yale faculty, he continued his research on generalized manifolds, publishing several important results, but he also took an interest in the improvement of undergraduate teaching, publishing an introductory calculus textbook in 1954 that took a fresh approach to a subject made dreary by generations of "cookbook" textbooks.

In 1951 Begle was elected Secretary of the American Mathematical Society, a position that brought him into the mainstream of the American mathematics community. He held the position for six years during a time when the Society was coping with problems brought on by the postwar expansion of interest and activity in mathematics.

Nationwide concern over the increased need for mathematicians led in 1958 to two conferences sponsored by the National Science Foundation, both of which called for a project to revamp the school mathematics curriculum. Begle was offered and accepted the post of director of the resulting project: The School Mathematics Study Group. He served as Director of SMSG for the duration of the project, from 1958 to 1972.

In 1961 Begle, and SMSG, moved to Stanford University, where Begle accepted a joint appointment as Professor in the Department of Mathematics and the School of Education. Ultimately, however, his appointment was shifted entirely to the School of Education, a shift that symbolized his growing interest in educational topics such as curriculum development and evaluation, teacher education, and research in mathematics education.

SMSG became the largest and most influential of the so-called "new math" curriculum projects. As its director, Begle organized teams of school teachers and mathematicians to prepare sample mathematics textbooks for each school grade, including special textbooks for less able students and for gifted students. He also set up panels and committees to develop supplementary books and programmed textbooks for students; monographs, textbooks, and films for teachers; and Spanish translations of many of these materials. He organized conferences to consider the nature and direction of future changes in school mathematics, and he sponsored research and evaluation studies of the effects of SMSG's curriculum development work.

At Stanford, Begle was active in the School of Education, especially after concluding his work with SMSG. He was Chairman of the Committee on Curriculum and Teacher Education from 1969 to 1975, Director of Teacher Education and Director of the Secondary Teacher Education Project from 1972 to 1975, and member of the Executive Board of the Stanford Center for Research and Development in Teaching in 1975. He was a member of the School of Education's Committee on Academic Affairs from 1972 to 1977, serving as Chairman from 1976 to 1977. He was a consultant to the California Statewide

Mathematics Advisory Committee from 1965 to 1966 and assisted the California State Department of Education in its test development efforts in the late 1960's.

Begle served on the Board of Directors of the National Council of Teachers of Mathematics from 1961 to 1964. From 1968 to 1973 he was on the Council's Finance Committee, and from 1974 to 1977 he was on its Research Advisory Committee. He was a member of the Committee on the Undergraduate Program in Mathematics of the Mathematical Association of America from 1959 to 1963 and an ex-officio member thereafter. He was especially active in the Committee's Panel on Teacher Training.

Begle served as Trustee of the American Mathematical Society from 1962 to 1967 and as Chairman of the Conference Board of the Mathematical Sciences from 1967 to 1969. He served two terms on the United States Commission on Mathematical Instruction, from 1962 to 1966 and from 1970 to 1975, and was Chairman from 1963 to 1966. In 1960 he was elected Fellow of the American Association for the Advancement of Science, and in 1975 he was appointed to the Executive Committee of the International Commission on Mathematical Instruction. In 1969 he was awarded the Distinguished Service Award of the Mathematical Association of America for his service to the mathematical interests of the nation. In 1971 he received the Rosenberger Medal from the University of Chicago for his significant contributions to humanity.

He died on 2 March 1978 in Palo Alto, California. He was survived by his wife, six of his seven children, and seven grandchildren.

Ed Begle was survived, too, by a mathematical community greatly in his debt for the leadership he provided during a critical period as mathematicians rediscovered the school. In his own career and through his influence on others, he worked to bridge the gulf between the research mathematician and the teacher of school mathematics. By encouraging, sponsoring, and conducting high quality research, he helped establish mathematics education as a respectable field in which serious scholarship was possible. He enriched the meaning of the term "mathematics educator."

Perhaps his greatest genius lay in his ability to discern what needed to be done and then to select and inspire other people to

join him in doing it. People responded to his call because of his
strong convictions, his great integrity and authority, and his
passion for the first-rate. He was a leader who saw his role as
service to others. He was an educator who taught by inspiring
his students to learn and to serve. He was an academic who,
looking beyond the walls of the academy and into school classrooms,
saw children struggling with pointless riddles and resolved to
make a change.

In his article, "The Math Wars," Benjamin DeMott, writing about
the members of the School Mathematics Study Group, observed that they

> chose not to isolate themselves in tight graduate school
> hives; they were willing to expose their studies and their
> present modes of thought to the eyes and understandings
> of humbler teachers and humbler students; they have not
> hesitated to accept responsibilities that other profes-
> sionals too often cynically delegate to publishers'
> drummers or third-rate academic minds. Even if they had
> failed in the task they undertook, they would for these
> reasons deserve praise and honor rather than abuse.
> As it is, they have a claim to a tolerably high rank
> among the exemplary intellectuals of this age.

Ed Begle, as a scholar and as a human being, was of the highest rank.

Publications

Many of E. G. Begle's writings, especially in later years, were anonymous or unpublished. The following list of references contains only his signed published work.

(with W. L. Ayres). On Hildebrant's example of a function without a finite derivative. *American Mathematical Monthly*, 1936, *43*, 294-296.

Locally connected spaces and generalized manifolds. *American Journal of Mathematics*, 1942, *54*, 553-574.

Intersections of contractible polyhedra. *Bulletin of the American Mathematical Society*, 1943, *49*, 386-387.

Regular convergence. *Duke Mathematical Journal*, 1944, *11*, 441-450.

Duality theorems for generalized manifolds. *American Journal of Mathematics*, 1945, *67*, 59-70.

A note on local connectivity. *Bulletin of the American Mathematical Society*, 1948, *54*, 147-148.

Topological groups and generalized manifolds. *Bulletin of the American Mathematical Society*, 1948, *54*, 969-976.

A note on S-spaces. *Bulletin of the American Mathematical Society*, 1949, *55*, 577-579.

A fixed point theorem. *Annals of Mathematics*, 1950, *51*, 544-550.

The Vietoris mapping theorem for bicompact spaces. *Annals of Mathematics*, 1950, *51*, 534-543.

Introductory calculus, with analytic geometry. New York: Holt, Rinehart & Winston, 1954.

The Vietoris mapping theorem for bicompact spaces II. *Michigan Mathematical Journal*, 1956, *3*, 179-180.

The School Mathematics Study Group. *Mathematics Teacher*, 1958, *51*, 616-618.

Comments on a note on "variable." *Mathematics Teacher*, 1961, *54*, 173-174.

A study of mathematical abilities. *American Mathematical Monthly*, 1962, *69*, 1000-1002; *Arithmetic Teacher*, 1962, *9*, 388-389; *Mathematics Teacher*, 1962, *55*, 648, 659.

Remarks on the memorandum "On the mathematics curriculum of the high school." *American Mathematical Monthly*, 1962, *69*, 425-426; *Mathematics Teacher*, 1962, *55*, 195-196.

The role of axiomatics and problem solving in mathematics. *American Mathematical Monthly*, 1967, *74*, 74-75.

Curriculum research in mathematics. In H. J. Klausmeier & G. T. O'Hearn (Eds.), *Research and development toward the improvement of education*. Madison, WI: Dunbar Educational Research Services, 1968.

SMSG: The first decade. *Mathematics Teacher*, 1968, *61*, 239-245.

The role of research in the improvement of mathematics education. *Educational Studies in Mathematics*, 1969, *2*, 232-244.

(with J. W. Wilson). Evaluation of mathematics programs. In E. G. Begle (Ed.), *Mathematics education* (Sixty-ninth Yearbook of the National Society for the Study of Education, Part 1). Chicago: University of Chicago Press, 1970.

Research and evaluation in mathematics education. In School Mathematics Study Group, *Report on a conference on responsibilities for school mathematics in the 70's*. Stanford, CA: SMSG, 1971.

SMSG: Where we are today. In E. W. Eisner (Ed.), *Confronting curriculum reform*. Boston: Little, Brown, 1971.

Time devoted to instruction and student achievement. *Educational Studies in Mathematics*, 1971, *4*, 220-224.

Teacher knowledge and student achievement in algebra (SMSG Reports No. 9). Stanford, CA: School Mathematics Study Group, 1972.

(with W. E. Geeslin). *Teacher effectiveness in mathematics instruction* (NLSMA Reports No. 28). Stanford, CA: School Mathematics Study Group, 1972.

(with L. B. Williams). *Calculus* (2nd ed.). New York: Holt, Rinehart & Winston, 1972.

The prediction of mathematics achievement (NLSMA Reports No. 27). Stanford, CA: School Mathematics Study Group, 1972.

Some lessons learned by SMSG. *Mathematics Teacher*, 1973, *66*, 207-214.

What's all the controversy about? (Review of *Why Johnny can't add* by M. Kline). *National Elementary Principal*, 1974, *53*(2), 26-31.

Basic skills in mathematics. In National Institute of Education, *Conference on Basic Mathematical Skills and Learning* (Vol. 1). Washington, DC: NIE, 1975.

The mathematics of the elementary school. New York: McGraw-Hill, 1975.

INJUSTICE

GODS AMONG US: YEAR FOUR

VOLUME 1

Jim Chadwick Editor – Original Series
David Piña Assistant Editor – Original Series
Jeb Woodard Group Editor – Collected Editions
Paul Santos Editor – Collected Edition
Steve Cook Design Director – Books
Louis Prandi Publication Design

Bob Harras Senior VP – Editor-in-Chief, DC Comics

Diane Nelson President
Dan DiDio and Jim Lee Co-Publishers
Geoff Johns Chief Creative Officer
Amit Desai Senior VP – Marketing & Global Franchise Management
Nairi Gardiner Senior VP – Finance
Sam Ades VP – Digital Marketing
Bobbie Chase VP – Talent Development
Mark Chiarello Senior VP – Art, Design & Collected Editions
John Cunningham VP – Content Strategy
Anne DePies VP – Strategy Planning & Reporting
Don Falletti VP – Manufacturing Operations
Lawrence Ganem VP – Editorial Administration & Talent Relations
Alison Gill Senior VP – Manufacturing & Operations
Hank Kanalz enior VP – Editorial Strategy & Administration
Jay Kogan VP – Legal Affairs
Derek Maddalena Senior VP – Sales & Business Development
Jack Mahan VP – Business Affairs
Dan Miron VP – Sales Planning & Trade Development

INJUSTICE: GODS AMONG US: YEAR FOUR VOLUME 1

Published by DC Comics. Cover and compilation Copyright © 2016 DC Comics. All Rights Reserved.

Originally published in single magazine form in INJUSTICE: GODS AMONG US: YEAR FOUR 1-7. Copyright © 2015 DC Comics. All Rights Reserved. All characters, their distinctive likenesses and related elements featured in this publication are trademarks of DC Comics. The stories, characters and incidents featured in this publication are entirely fictional. DC Comics does not read or accept unsolicited ideas, stories or artwork.

DC Comics, 2900 West Alameda Ave., Burbank, CA 91505
Printed by RR Donnelley, Salem, VA, USA. 3/25/16. First Printing.

ISBN: 978-1-4012-6130-6

Library of Congress Cataloging-in-Publication Data is Available.

THE STORY SO FAR

This is not the world as you know it. This is a world where the Joker destroyed Metropolis in an atomic attack that claimed the lives of Lois Lane and her unborn child with Superman. This is a world where the Man of Steel, mad with grief, murdered the Joker in cold blood as Batman looked on in horror.

From that moment on, everything changed. Superman started going further and further to bring justice to the entire world, even involving himself in civil wars. As Batman became concerned about Superman's increasing global power, the Justice League found themselves split between members loyal to Superman and those who shared Batman's concerns.

Soon, Superman's team came into conflict with the U.S. government, and the battle between the two former friends started to see casualties mount on each side. Batman's resistance team stole a Kryptonite-powered pill that grants the user superpowers, to even the fight between Superman's super-powered troops and Batman's allies in the Gotham City Police Department.

Meanwhile, Superman found himself with an unexpected new ally: Sinestro, the former rogue Green Lantern who formed his own fear-powered Sinestro Corps. Sinestro recruited both Superman and Hal Jordan into the Sinestro Corps, and a huge war with the Green Lantern Corps and Batman's resistance followed, with casualties on both sides. Batman's team captured Raven, Flash and Robin from Superman's side and put Wonder Woman into a magical coma, but, despite these setbacks, Superman now seemed unstoppable.

To combat a Superman empowered with a yellow power ring, Batman turned to the one force he knew can damage Superman—magic. Teaming up with John Constantine, who wanted revenge on Superman for causing the events that lead to the death of his daughter's mother, Batman gathered an impressive array of magic-user to his side and took to the Tower of Fate to plan his attack on Superman.

But Superman found himself with an unlikely magical ally—The Spectre, the Spirit of Vengeance. With the Spectre on his side, Batman's team couldn't dare attempt an assault on Superman. But as the Spectre began acting erratically, killing powerful beings like Deadman and the demon Etrigan's human host Jason Blood, Batman and Constantine hatched a plan to counter him. While Batman became the new human host for Etrigan, Constantine tricked the demon Trigon into thinking Superman, not Batman, had kidnapped Trigon's daughter Raven.

But all was not as it seemed. The Spectre had been controlled by Mr. Mxyptlk, the fifth-dimensional being who was obsessed with Superman. As the war between Mxyptlk and Trigon threatened to unravel reality, a desperate final plan—executed with the help of Dick Grayson, now the new Deadman—banished both Mxyptlk and Trigon to another dimension, but left Batman without his new magical allies.

Now, Batman is hidden deep underground, his team reduced to just a handful of stalwarts. And Superman's team is reunited—Flash, Raven and Robin were rescued, and Wonder Woman was awoken from her coma when her mother Hippolyta made a deal with the goddess Hera. With Diana back by his side, Superman rejected Sinestro's ring.

Times have never been brighter for Superman's team, or looked darker for Batman. But in this world, things have a habit of changing quickly...

"The Gods Themselves" **Bruno Redondo** Penciller **Juan Albarran** Inker **Rex Lokus** Colorist
Cover Art by **Howard Porter & Rex Lokus**

THE GODS THEMSELVES

OLYMPUS.

A KINGDOM THAT A SCANT FEW BELIEVE EXISTS...AND EVEN FEWER HAVE EVER SEEN.

MOST ACCEPT ONLY THE WORLD THEY SEE, AND ARE OBLIVIOUS TO THE ALTERNATE TIMELINES, MAGICAL DIMENSIONS, AND PLACES THAT EXIST OUTSIDE OF KNOWN REALITY.

BUT OLYMPUS IS NOT MERELY FOLKLORE OR MYTH MEANT TO ACCOUNT FOR WHAT COULD NOT BE EXPLAINED BY PRIMITIVE MAN.

OLYMPUS IS REAL, AND SO ARE THE GODS THAT INHABIT IT. WE ARE A POWERFUL AND PASSIONATE LOT, AND EACH OF US HAVE UNIQUE ROLES TO PLAY. THERE ARE GODS FOR MARRIAGE, WISDOM, LOVE, ARCHERY...AND EVEN WINE.

I AM ARES... THE GOD OF WAR.

I ENVY YOUR TRIPS TO THE MORTAL WORLD...

...BUT I MADE A PROMISE TO MY FATHER THAT I WOULD NOT RETURN TO THE WORLD OF MAN UNLESS BY HIS DECREE.

THAT SAID, I DO MISS THE EXCITEMENT AND GLORY OF MAN'S WORLD. IT CAN BE SO BORING AMONG THE GODS.

POOR, SIMPLE HERCULES...

EARTH HAS UNDERGONE A DRAMATIC TRANSFORMATION OVER THE LAST THREE YEARS. THESE PAST MONTHS HAVE BEEN THE MOST PEACEFUL TIME THIS PLANET HAS EVER SEEN.

BUT THE COST OF PEACE HAS BEEN ENORMOUS...AND HAS TAKEN A TOLL ON EVERYONE IN THE JUSTICE LEAGUE.

NONE MORE THAN SUPERMAN.

HE LOST HIS WIFE, HIS UNBORN SON, AND HIS CITY. AND ALONG WITH THAT, A PIECE OF HIS HUMANITY.

HE ALSO LOST HIS BEST FRIEND... A BETRAYAL HE WON'T LET GO.

THE INSURGENCY HAS BEEN DORMANT SINCE OUR BATTLE WITH TRIGON...BUT THAT HASN'T KEPT SUPERMAN FROM HAVING THE JUSTICE LEAGUE SCOUR THE EARTH LOOKING FOR BATMAN.

IT'S LIKE HE WON'T GIVE UP UNTIL HE ACTUALLY HEARS FROM BRUCE'S MOUTH THAT IT'S OVER.

NOW, I'VE BEEN ACROSS THE UNIVERSE AND HAVE SEEN THE WILDEST, MOST MIND-BLOWING THINGS IMAGINABLE...

...BUT BATMAN GIVING UP? THAT'S JUST CRAZY.

HE DIDN'T JUST FALL OFF THE FACE OF THE PLANET, HAL...

THE PACIFIC LEAD MINES, NORTHWEST ALASKA.

"SIR, AT THE RISK OF SOUNDING RIDICULOUS...

...ARE YOU AWARE THAT THOSE SCREENS ARE BLANK?

YES, ALFRED. I JUST TURNED THEM OFF.

EXCELLENT. THEN THIS IS THE PERFECT OPPORTUNITY FOR YOU TO TAKE A MEAL.

THANK YOU, BUT I'M NOT INTERESTED IN EATING--

AND I AM LESS INTERESTED IN RECITING THE BENEFITS OF GOOD NUTRITION, MASTER BRUCE...

FINE.

ALFRED...ARE YOU SERIOUSLY GOING TO STAND OVER ME TO MAKE SURE I EAT?

"Vengeance Is Mine" Mike S. Miller Artist J. Nanjan Colorist
Cover Art by Art Thibert & Thomas Mason

VENGEANCE IS MINE

WHAT ARE YOU DOING?

PLAYING HOOKY. HOP ON.

I'M NOT GETTING ON THAT THING.

IT'S NOT A THING. IT'S A HARLEY. GET IT... A *HARLEY*?

YEAH. I GOT IT. STILL NOT GETTING ON.

OKAY, THEN... DO THE SWITCHEROO! CHANGE THOSE PIMPLES TO DIMPLES AND FLY US OUTTA HERE.

...MPLES TO PIMPLES?

YOU KNOW, YOUR BETTER HALF-- THE ONE WITH ALL THE MUSCLES.

YOU EXPECT ME TO CHANGE AND FLY AWAY WITH YOU?

UM... YEAH.

NO.

THOK

YOU SUCK.

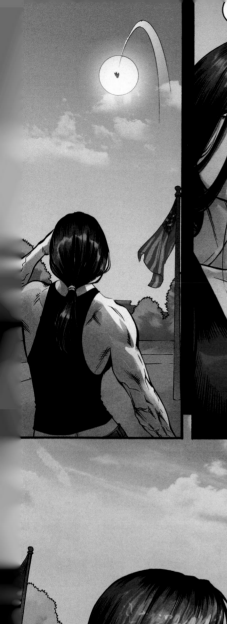

"Choices" **Bruno Redondo** Penciller **Juan Albarran** Inker **Rex Lokus** Colorist
Cover Art by **Tom Raney & Thomas Mason**

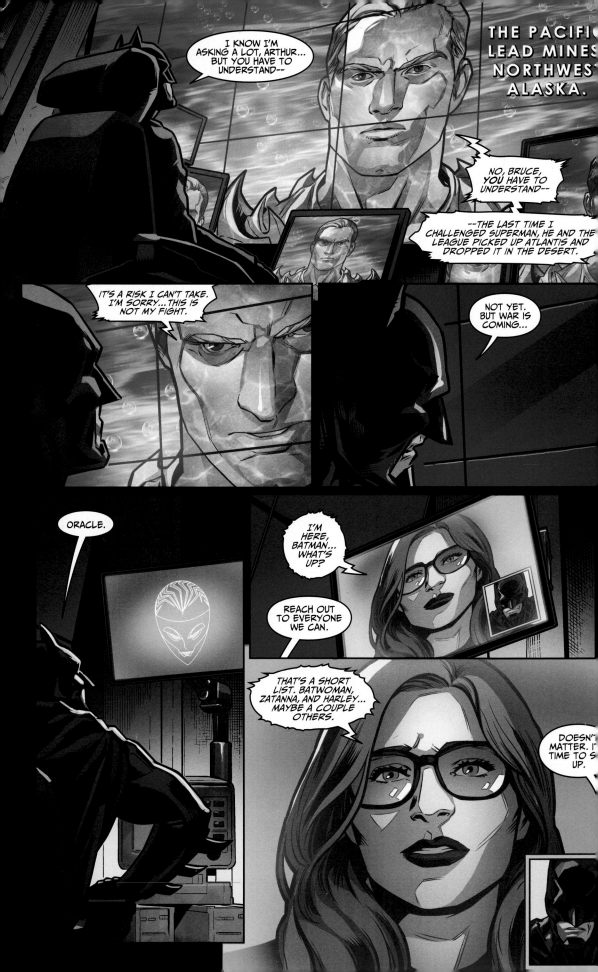

"Bargins" Mike S. Miller Artist J. Nanjan Colorist
Cover Art by Yildiray Cinar & Rex Lokus

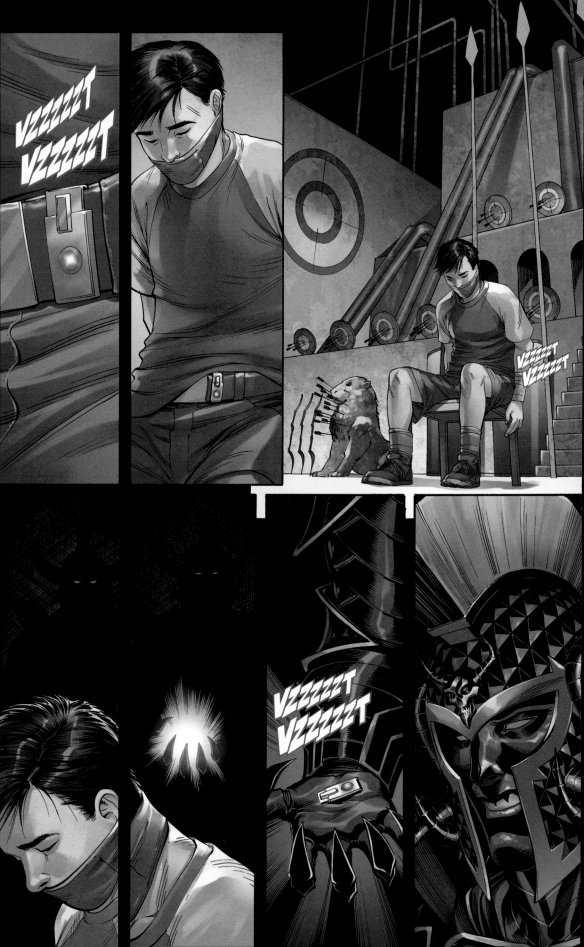

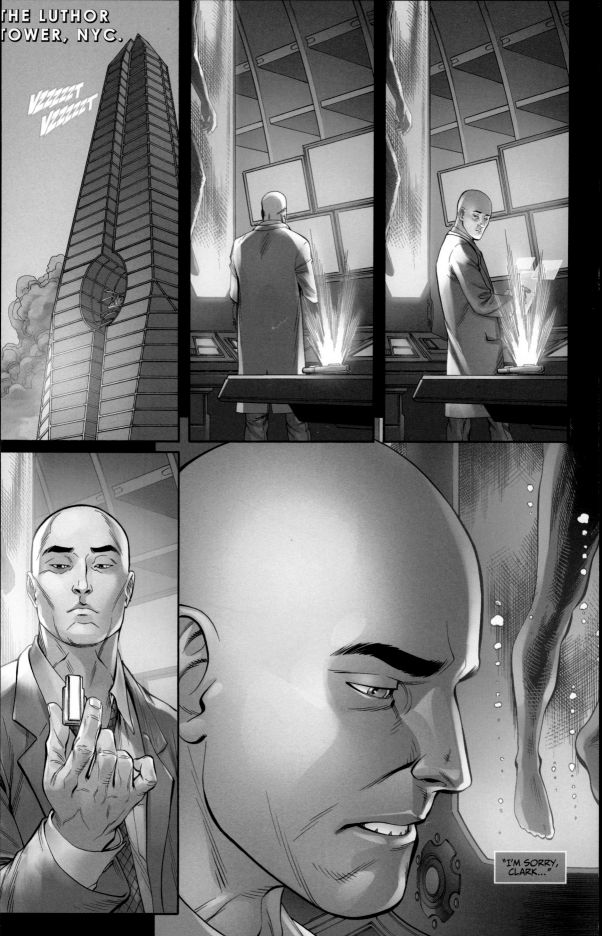

I FOUGHT AS AN AMAZON IN TRIAL BY COMBAT--

--NOW YOU DARE ASK ME TO STAND WITH THE INSURGENCY IN THE NAME OF ZEUS?!

I WILL NOT!

THERE IS *NO* DECEPTION HERE. I AM HIS MESSENGER. MY WORDS ARE HIS WORDS--

LET HIM COME BACK AND TELL ME HIMSELF!

DIANA, STOP!

SHOULD WE JUMP IN?

IT ENDS NOW,
FEARMONGER...

KWHUMP

"Strength of Hercules" Bruno Redondo Penciller Juan Albarran Inker Rex Lokus Colorist
"The Old and the New" Xermanico & Tom Derenick Artists Rex Lokus Colorist
Cover Art by **Jae Lee & June Chun**

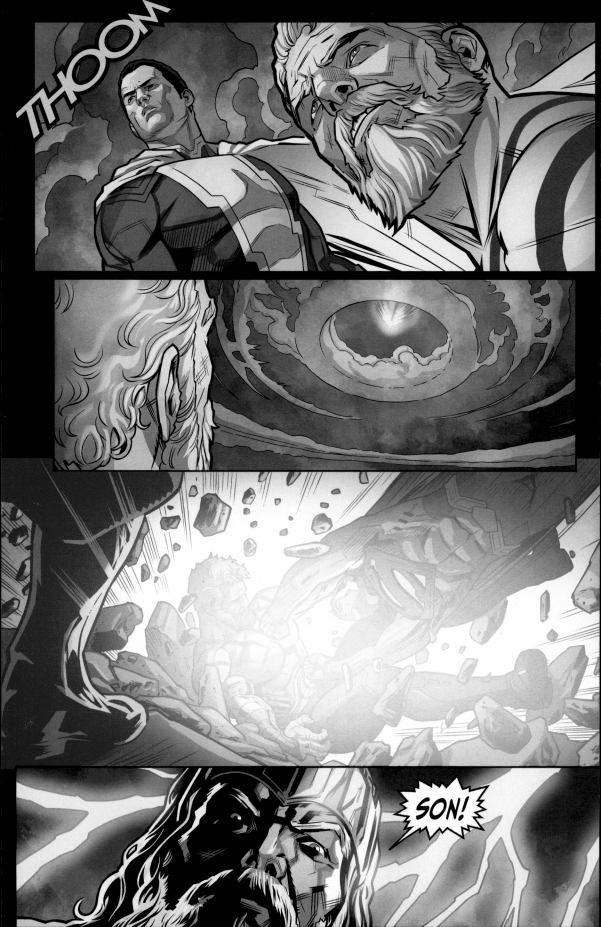